Ground-Penetrating Radar for Archaeology

Geophysical Methods for Archaeology

Series Editors:
Lawrence B. Conyers, University of Denver,
Kenneth L. Kvamme, University of Arkansas

The Geophysical Methods for Archaeology series aims to introduce field archaeologists and their students to the theory and methods associated with near-surface geophysical data collection techniques. Each book in this series will describe one of these commonly used noninvasive techniques, its applications, and its importance to archaeological practice for the non-specialist.

Volume 1: *Ground-Penetrating Radar for Archaeology*, Lawrence B. Conyers

Ground-Penetrating Radar for Archaeology

LAWRENCE B. CONYERS

ALTAMIRA
PRESS

A Division of
ROWMAN & LITTLEFIELD PUBLISHERS, INC.
Walnut Creek • Lanham • New York • Toronto • Oxford

PRESS

A division of Rowman & Littlefield Publishers, Inc.
1630 North Main Street, #367
Walnut Creek, California 94596
www.altamirapress.com

Rowman & Littlefield Publishers, Inc.
A wholly owned subsidiary of The Rowman & Littlefield Publishing Group, Inc.
4501 Forbes Boulevard, Suite 200
Lanham, MD 20706

PO Box 317
Oxford
OX2 9RU, UK

Copyright © 2004 by AltaMira Press

All rights reserved. No part of this publication may be reproduced, stored in a retrieval system, or transmitted in any form or by any means, electronic, mechanical, photocopying, recording, or otherwise, without the prior permission of the publisher.

British Library Cataloguing in Publication Information Available

Library of Congress Cataloging-in-Publication Data

Conyers, Lawrence B.
 Ground-penetrating radar for archaeology / Lawrence B. Conyers.
   p. cm.
 Includes bibliographical references and index.
 ISBN 0-7591-0772-6 (hardcover : alk. paper) — ISBN 0-7591-0773-4 (pbk. : alk. paper)
 1. Geophysics in archaeology. 2. Ground penetrating radar. I. Title.

CC79.G46C663 2004
930.1'01'55—dc22
                                                                    2004009506

Printed in the United States of America

∞™ The paper used in this publication meets the minimum requirements of American National Standard for Information Sciences—Permanence of Paper for Printed Library Materials, ANSI/NISO Z39.48-1992.

Contents

List of Illustrations   ix

Acknowledgments   xiii

1 Archaeological Geophysics   1
Geophysical Methods for Archaeological Mapping   3
History of Geophysical Surveys in Archaeology   4
Archaeological Geophysics Today and the Goals of This Book   6

2 Introduction to Ground-Penetrating Radar   11
History of GPR in Archaeology   16

3 GPR Theory and Practice   23
Radar Energy Generation and Propagation   23
Generation and Recording of GPR Waves   26
Acquisition Procedures   28
Collection of Transects in Grids   34
Data Recording   37
Antenna Variables   39
Radar Propagation and Reflection in the Ground   45
Measurements of Radar Propagation and Reflection ▪ *Dispersion and Attenuation of Radar Energy in the Ground* ▪ *Reflection Types* ▪ *Resolution of Subsurface Features*

Radar Propagation and Reflection Complications 68
Ground Coupling ▪ *Background Noise* ▪ *Focusing and Scattering Effects* ▪ *The Near-Field Effect* ▪ *Air Waves and Near-Surface Obstructions*

4 Ground-Penetrating Radar Equipment and Acquisition Software 81
GPR Systems 81
Data Acquisition Software: Setup Parameters 84
Header Information ▪ *Time Window* ▪ *Samples per Reflection Trace* ▪ *Trace Stacking* ▪ *Transmission Rate* ▪ *Time Zero Position* ▪ *Range Gains* ▪ *Vertical Filters*

5 Velocity Analysis 99
Reflected Wave Methods 102
Direct-Wave Methods 105
CMP and WARR Tests ▪ *Transillumination Tests*
Laboratory Measurements of RDP 112
Analysis of Point Source Reflection Hyperbolas 115
Velocity Analyses Conclusions 116

6 Postacquisition Data Processing 119
Scale Correction and the Creation of Reflection Profiles 120
Remove Horizontal Banding 123
Removal of High-Frequency Noise 125
Removal of Multiple Reflections 126
Migration 128
Increase the Visibility of Subtle Reflections 129
Data Processing Conclusions 131

7 Interpretation of GPR Data 133
Synthetic GPR Models 134
Creating a Synthetic Computer Model ▪ *Synthetic Modeling Applications* ▪ *Synthetic Models Compared to GPR Profiles*
Interpreting and Mapping Many GPR Reflection Profiles in a Grid 144
Example of Buried Landscape Reconstruction from Interpreting Reflection Profiles
Amplitude Analysis in Slice Maps 148
Amplitude Slice Maps on Level Ground ▪ *Amplitude Slices on Uneven Ground* ▪ *Subtle Feature Discovery with Amplitude*

Mapping ▪ *Amplitude Maps to Search for Vertical Features and Graves* ▪ *Production of Rendered Images*

8	**Conclusion**	165
	References	175
	Index	197
	About the Author	203

List of Illustrations

FIGURES

Figure 2.1.	A GPR Reflection Profile	12
Figure 2.2.	GPR Equipment	14
Figure 3.1.	Electromagnetic Wave Propagation	24
Figure 3.2.	The GPR Frequency Distribution	25
Figure 3.3.	The Generation of a Waveform	26
Figure 3.4.	GPR Antennas Collecting Data	27
Figure 3.5.	An Antenna Survey Wheel	29
Figure 3.6.	A Reflection Trace	30
Figure 3.7.	A One-Person GPR System	31
Figure 3.8.	Topographically Corrected Reflection Profile	33
Figure 3.9.	Antenna Frequency Distribution	40
Figure 3.10.	Radar Pulses	41
Figure 3.11.	Low-Frequency Antennas	42
Figure 3.12.	High-Frequency Antennas	43
Figure 3.13.	Graph of the Relative Dielectric Permittivity–Velocity Relationship	46
Figure 3.14.	Reflection Profile in Wet Clay	51
Figure 3.15.	Point Source Hyperbolas	54
Figure 3.16.	Many Small Hyperbolas	56
Figure 3.17.	Generation of a Reflection Hyperbola	57
Figure 3.18.	Conical Spreading of Radar Energy in the Ground	62

Figure 3.19.	Radiation Footprint Differences with Differing Ground RDP	63
Figure 3.20.	Resolution of Interfaces	65
Figure 3.21.	Resolution of Stratigraphy as a Function of Frequency	66
Figure 3.22.	Energy Focusing with Depth	69
Figure 3.23.	Coupling Changes Due to Differences in Surface Materials	70
Figure 3.24.	Coupling Changes Producing Anomalous Reflections	71
Figure 3.25.	Extreme Electromagnetic Noise in a Reflection Profile	72
Figure 3.26.	Energy Scattering and Focusing	74
Figure 3.27.	Scattering and Focusing on a Horizontal Reflection Surface	75
Figure 3.28.	Buried Agricultural Field That Causes Focusing and Scattering	75
Figure 3.29.	Air Waves	77
Figure 3.30.	Near-Surface Metal Interference	78
Figure 4.1.	Range Gaining	92
Figure 4.2.	Trace Clipping	94
Figure 4.3.	Frequency Filtering	97
Figure 5.1.	Stratigraphy Identification for Velocity Determination	102
Figure 5.2.	Travel Paths for a CMP Test	106
Figure 5.3.	Conducting a CMP Test	107
Figure 5.4.	A CMP Test	108
Figure 5.5.	Conducting a Transillumination Test	111
Figure 5.6.	Transillumination Test Results	113
Figure 5.7.	Laboratory Tests of RDP	115
Figure 5.8.	Hyperbola-Fitting Velocity Tests	117
Figure 6.1.	Wiggle Trace and Gray-Scale Reflection Profiles	121
Figure 6.2.	Raw Reflection Profile and Adjusted Profile	122
Figure 6.3.	Background Removal Processing	124
Figure 6.4.	Multiple Reflections	127
Figure 6.5.	Reflection Migration	130
Figure 7.1.	Possible Ray Paths in Stratified Material	137
Figure 7.2.	A V-Trench Synthetic Model	139
Figure 7.3.	A V-Trench Discovered in a Reflection Profile	140
Figure 7.4.	A Three-Layer Velocity Model	141

LIST OF ILLUSTRATIONS

Figure 7.5.	Vertical Distortion Due to Near-Surface Velocity Differences	142
Figure 7.6.	A Pit House Floor with Possible Subfloor Feature	143
Figure 7.7.	Synthetic Model of a Pit House Floor	144
Figure 7.8.	Hand Mapping of Reflection Profiles in a Grid	147
Figure 7.9.	Slicing Choices in Topographically Complex Areas	155
Figure 7.10.	Reflection Profile with No Distinctive Reflections	157
Figure 7.11.	Stratigraphic Truncation in a Cemetery	160
Figure 7.12.	Graves with Distinct Reflection Hyperbolas	161
Figure 7.13.	Three-Dimensional Rendered Surface	163

TABLES

Table 3.1.	Typical Relative Dielectric Permittivities (RDPs) of Common Geological Materials	47
Table 3.2.	Wavelength of Radar Waves in Media of a Given RDP and Frequency	60
Table 4.1.	Depth in Meters to a Reflector through Media of a Given Relative Dielectric Permittivity	86
Table 6.1.	Common Postacquisition Processing Objectives and Methods	120
Table 8.1.	Feasibility of Using GPR to Discover and Map Some Buried Archaeological Features and Stratigraphy	169

PLATES

Plate 1.	Three-Dimensional View of a Buried Living Surface
Plate 2.	Amplitude Slice Maps
Plate 3.	Slices Crossing Subsurface Bedding Planes
Plate 4.	Anomalies Created by Slices Crossing Bedding Planes
Plate 5.	Interpolation Differences in Slicing
Plate 6.	Amplitude Slices on Horizontal Ground
Plate 7.	Amplitude Slices on a Mound
Plate 8.	Amplitude Slices Illustrating Subtle Ground Features
Plate 9.	Grave Amplitude Slice Map
Plate 10.	Amplitude Slice Maps of a Very Subtle Floor Feature

Acknowledgments

Archaeological geophysics is not the kind of research one can do alone. Every project I have worked on, whether field research, data processing, or software development, has involved someone in a crucial capacity to whom I owe a great deal. My great thanks go to Jeffrey Lucius at the U.S. Geological Survey in Denver, Colorado, who first held my hand and taught me the basics of GPR when I was trying to get my feet on the ground in near-surface geophysics many years ago. He has always been there to answer questions about electromagnetic theory, help with data processing, and recently assist in the writing of new computer programs for GPR analysis. A more patient and giving colleague would be hard to find in any scientific endeavor. Dean Goodman, with whom I coauthored the first GPR book for archaeology, *Ground-Penetrating Radar: An Introduction for Archaeologists*, has also been a constant inspiration. He consistently amazes me with his grasp of the complexities of GPR and especially with his ability to write computer code for processing reflection data.

My geophysical and archaeological colleagues who continue to believe in geophysics through both our successes and failures are the intellectual capital on which the future of archaeological geophysics will be built. Great thanks go to the following people who have helped, collaborated with, encouraged, and advised me over the years: John Hildebrand, Sean Wiggins, Tom Carr, Payson Sheets, Ken Kvamme, Cathy Cameron, John Isaacson, Mike Hargrave, Steve DeVore, Mike Powers, Jay Johnson, Clark Davenport, Doug Wilson, Leigh-Ann Bedal, David

Hurst Thomas, Karyn deDufour, Terry Ferguson, Mark Gilberg, Jeffrey Quilter, Jon Kent, Jim Thompson, and Yasushi Nishimura.

Little research in an academic setting could be accomplished without students, who not only help in data collection and join the intellectual excitement of discovery, but perform much of the everyday data processing and map making that is such a time-consuming but important part of this research. They are more than students but colleagues, always asking the pertinent questions and posing the important problems that must be answered. It is often they who make many of the cumulative breakthroughs that allow the science of archaeological geophysics to progress. Many thanks go to my students, many of whom have moved on to careers in geophysical archaeology, including Eileen Ernenwein, Derek Hamilton, Michael Grealy, Tiffany Osburn, Kimberley Henderson, Tiffany Tchakirides, Michele Koons, and Jennie Sturm.

Funding for a good deal of the field and laboratory research presented in this book was provided by the National Center for Preservation Technology and Training, National Geographic Society, the Strategic Environmental Research and Development Program of the U.S. Department of Defense, Environmental Protection Agency and Department of Energy, the Colorado Historical Society, and the U.S. National Park Service. The private cultural research management firms, individuals, corporations, and other organizations that sometimes actually pay good money to us for doing GPR on their projects have also been an important funding mechanism, and many thanks go to them for continuing to believe in and support geophysical archaeology.

1

Archaeological Geophysics

Archaeological geophysics is a method of data collection that allows field archaeologists to discover and map buried archaeological features in ways not possible using traditional field methods. Using a variety of instruments, physical and chemical changes in the ground related to the presence or absence of buried materials of interest can be measured and mapped. When these changes can be related to certain aspects of archaeological sites such as architecture, use areas, or other associated cultural features, high-definition maps and images of buried remains can be produced. Geophysical techniques are usually most effective at buried sites where artifacts and features of interest are located within 2 to 3 meters of the surface, but they have occasionally been used for more deeply buried deposits.

A small but growing community of archaeologists have been incorporating geophysical mapping techniques as a routine field procedure for many years (Gaffney and Gater 2003). Their maps act as primary data that can be used to guide the placement of excavations or to define sensitive areas containing cultural remains to avoid. Some archaeological geophysicists have also started using these methods as a way to place archaeological sites within a broader environmental context to study human interaction with and adaptation to ancient landscapes (Kvamme 2003).

Ground-penetrating radar (GPR) is one of these geophysical methods that involves the transmission of high-frequency radar pulses from a surface antenna into the ground. The elapsed time between when this energy is transmitted, reflected from buried materials or sediment and soil changes in the ground, and then received back at the surface is measured. When many thousands of radar reflections are measured and recorded as antennas are moved along transects within a grid, a three-dimensional picture of soil, sediment, and feature changes can be created.

Mapping using GPR as well as other geophysical techniques has recently become so accurate that the possibility now exists to test any number of working hypotheses concerning a broad range of anthropological, geological, and environmental questions. Some of those could be related to social organization and social change, when these cultural attributes can be directly related to the placement, orientation, size, geometry, or distribution of certain architectural and ancillary features on the landscape. Determining geological and environmental aspects of ancient landscapes such as soil changes and the nature of buried topographic features is also possible (Conyers 1995; Conyers and Spetzler 2002; Conyers et al. 2002). Most important, GPR and other geophysical methods can gather a great deal of information about the near surface in a totally nondestructive way, allowing large areas with buried remains to be studied efficiently and accurately, while at the same time preserving and protecting them.

Recent advances in all near-surface geophysical methods, including GPR, have captured the imagination of many in the archeological community and have prompted some field archaeologists to routinely employ geophysical techniques in their own research. Although only a small fraction of field archaeologists, especially in the Western Hemisphere, are routinely including GPR and other methods as part of their "bag of tricks," the field is growing rapidly and will soon become, one hopes, a standard and accepted operating field procedure. Cultural resource managers have rapidly grasped the power of geophysical methods to quickly, efficiently, and nondestructively discover and map sites for selective excavation or avoidance, producing greater economy of time and resources (Johnson 2004). Research archaeologists have also been drawn to the power of geophysics not only to discover buried remains but also to place subsurface information from standard excavations into an overall site context. In this way, a limited amount of information from the known can be projected to areas of a site that remain buried and often will remain so. Ultimately, geophysical mapping allows for a more complete analysis of many archaeological sites in ways that could only be dreamed about just a few years ago because of its ability to evaluate large areas of buried and otherwise invisible archaeological sites quickly and accurately.

The goal of this book is to introduce all types of archaeological researchers to the power of GPR and to inform and guide those who hope to use, or have already used, these techniques in their work. This will be done by discussing the most commonly used data collection and processing methods, using case studies from around the world to illustrate both successes and failures, in order to demonstrate the power as well as pitfalls of the method. While these procedures can vary depending on the questions asked, local conditions encountered, equipment used, and the data collection, processing, and interpretation methods employed, most of the basic methods will be discussed.

GEOPHYSICAL METHODS FOR ARCHAEOLOGICAL MAPPING

The most common near-surface geophysical methods used in archaeology are magnetometry, resistivity, electromagnetic conductivity, and the topic of this book, GPR (Clark 1990; Gaffney and Gater 2003). Magnetic methods employ passive devices that measure small changes in the Earth's magnetic field that are influenced by changes in soils and buried materials below the surface. These changes, if related to cultural or geological phenomena of interest, are then mapped spatially and can tell much about the patterning of some magnetic features in the near surface. The other three common geophysical methods are active methods, in that they transmit energy into the ground and then measure how that energy is affected by cultural, geological, or environmental changes in the ground. As with magnetometry, it is hoped that the mapped changes can be related to phenomena of interest to the archaeologist, such as the presence or absence of buried cultural features or geological changes that are meaningful. Resistivity transmits an electrical current into the ground and measures the differences in voltage between the transmitting device and a recording device some distance away. A similar method of energy transmittal is used in electromagnetic (EM) conductivity, except a geometrically complex EM field is induced into the ground, and measurements are made of the effect of that field on the underlying deposits. In this method, both electrical and magnetic properties of the ground can be measured.

Many other geophysical techniques can be used to measure other properties of materials in the near surface, which are either not commonly used or still in the experimental stages. One of these is thermal imaging, which measures the radiation of heat from the ground over periods of time. Changes in radiation are theoretically related to differences in materials near the surface. Spontaneous potential measures the background electrical potential of buried materials, which differ based on their composition and water content (Reynolds 1998). Magnetic susceptibility is a technique that takes readings of the ground directly through boreholes or probes in order to measure the remnant magnetism of buried materials. These readings can often be related to soil property changes that are affected by human modification of the landscape, or natural soil and sediment formation processes. Seismic reflection and refraction is similar to ground-penetrating radar except the active energy source propagated into the ground is sonic waves and not radar energy. In the future, the seismic method has the potential to map buried sites in three dimensions much like GPR, but at present it is hampered somewhat by its expense and the slowness in collecting and processing data (Hildebrand et al. 2002; Ovenden 1994).

HISTORY OF GEOPHYSICAL SURVEYS IN ARCHAEOLOGY

Archaeologists have long experimented with methods that will allow them to see or visualize in some way what is below the ground, sometimes resorting to many relatively crude methods such as random or systematic shovel tests, trenching, soundings, probing, and even nose-sensitive dogs or other even less scientific techniques such as dowsing (van Leusen 1998). These approaches, for the most part, have proved to be less than accurate, often expensive, potentially destructive, or producing small looks at the subsurface that are not statistically representative. As a result, many deeply buried or otherwise invisible sites remained mostly hidden and unstudied. It is those buried sites that are most suitable for discovery and mapping using near-surface geophysics.

A few rudimentary geophysical surveys that attempted to map buried cultural remains were carried out in Europe and North America in the 1920s and 1930s, but these proved to be, at best, anomaly-generating exercises that were difficult to interpret (Gaffney and Gater 2003: 14). These early experiments were conducted primarily with magnetic and electrical tools developed for mining and petroleum exploration applications, and the anomalies recorded by them were often found to be related more to geological changes rather than to the presence of archaeological features in the ground.

Beginning in earnest during the 1950s, a few pioneering geophysicists began experimenting with electrical and magnetic methods as a way to quantify ground conditions and potentially discover and map hidden archaeological remains (Bevan 2000). In contrast with many of today's techniques, these initial attempts at "seeing below the soil" were crude but still effective enough to generate usable maps of buried sites, which piqued the interest of some archaeologists. Some of these early surveys were conducted with equipment as simple as a car battery, wires, and a voltmeter (Bevan 1998). Others were little more than sophisticated metal detectors, but ultimately these early studies (Aitken 1958) paved the way for all future work.

For the most part, early geophysical surveys collected data that were recorded as data points on paper, for later hand mapping. Some data were recorded on magnetic tape, which could later be digitized, but those more sophisticated data storage and processing attempts were generally the exception. Sometimes field data, as in the case of early ground-penetrating radar reflections, were printed on paper and could later be analyzed in three dimensions, but often this step was time-consuming and fraught with processing and interpretation problems (Bevan and Kenyon 1975). Having to write data points collected in the field on paper, or use paper copies of field records, and then attempt to make sense of them later on limited the amount of area that could be surveyed and the types and quality of data processing that could be accomplished.

With the advent of fast and relatively inexpensive computers in the mid-1980s, geophysical systems evolved rapidly into tools that could record data digitally on computer disks or tape, and then allow those data to be processed after returning from the field. This enhanced their quality and speeded mapping and subsequent interpretation. Collecting data digitally allowed much larger areas of ground to be covered and increased the data density dramatically, creating maps and images of much greater precision. The small but enthusiastic archaeological geophysical community quickly recognized the power of computer collection and processing, and near-surface geophysical techniques advanced rapidly throughout the 1990s.

Today, with some systems, many megabytes (sometimes hundreds of megabytes) of data can be recorded each day, covering large areas of land with very dense grids of potentially meaningful data. With the computer-processing power now available, these data can be readily made into usable maps, sometimes within hours of collection, giving geophysics for archaeology the "immediate gratification" component that was heretofore reserved for those archaeologists digging and observing artifacts and features in standard excavations. This ability to quickly produce accurate images of below ground features in a way that can be immediately interpreted not only is useful but also gives these methods a greater legitimacy with the more traditional archaeologists. It is this power to produce usable images of buried materials quickly and accurately that has transformed geophysics from what would otherwise appear to be strange squiggles on a computer screen, or streams of data on a computer hard drive, into a technique that produces understandable and immediately usable maps. Geophysical maps can then be readily interpreted by field archaeologists in order to plan excavations and understand the nature of buried deposits. Most important, the output is in a form that the human brain can readily interpret, which is one of the reasons that archaeological geophysics has recently seen such a surge of interest from the archaeological community.

ARCHAEOLOGICAL GEOPHYSICS TODAY AND THE GOALS OF THIS BOOK
Many techniques of archaeological geophysics used today have been borrowed or modified from other disciplines, making this subfield of archaeology, by necessity, multidisciplinary. Most of the early techniques that showed promise were developed by researchers with physics or geology backgrounds, and they were often applied to archaeological sites as an adjunct to other more important data-gathering tasks. Magnetometry was originally developed by researchers hoping to locate geological structures capable of containing economically valuable minerals or other deposits (Aitkin 1958). Some of those electromagnetic tools were developed to map deposits as varied as hazardous wastes, volcanic intrusions, and groundwater deposits (Reynolds 1998). Ground-penetrating radar as we know it today was originally developed for the U.S. space program to map the depth and variation of deposits on the moon (Simmons et al. 1972). The technique, as with many other geophysical methods, was quickly modified and adapted for many geotechnical applications and ultimately to archaeology.

Today most of the near-surface geophysical instruments are still manufactured for applications other than archaeology, as no manufacturer can afford to develop these expensive tools for the archaeological market alone. As a result, the archaeological community is almost always forced to use "off-the-shelf" geophysical systems, whose manufacturers are motivated by profit (little of which is generated from archaeological customers). The archaeological community by necessity has had to learn how to make these geophysical systems developed for other types of studies work for their own needs. This can be both a blessing and a curse. While it is nice to have geophysical systems that are well tested and supported by manufacturers that can be used for archaeological mapping, often the systems' standard data collection and processing procedures must be modified for archaeological needs. Also, much of the standard software processing and imaging packages were developed for scientists more interested in finding buried pipes or geological deposits, and therefore they must be diligently applied and often modified for archaeological requirements. Recently, by necessity, archaeological geophysicists have had to produce their own software programs specifically for archaeological applications.

Today almost all field archaeologists, at least in the Western Hemisphere, are trained in anthropology departments, although there are always a few students that make their way to archaeology from other disciplines. Those of us who teach in anthropology departments are quite aware that many of our students were drawn to archaeology for a number of nonscientific reasons, such as their joy of finding or working with interesting artifacts, perhaps their inability to get through advanced mathematics or science courses, and even the romance of archaeology in popular culture. In the recent past, this has often meant that many anthropology students had a very difficult time comprehending the physics or math that is necessary to understand geophysics, which may be one reason archaeological geophysics has grown relatively slowly compared to geophysics employed in other disciplines. This situation is now changing rapidly as a younger generation, raised with computers and not necessarily terrified by the prospect of analyzing digital data, has entered the field. It is encouraging that many of this new generation of archaeology students are ready both to learn and apply geophysical techniques in their own research, but unfortunately many are still not introduced to the subject in their typical college class work. This situation can only improve as geophysical methods become more common and are shown to be successful, and as a new generation of computer-savvy students move into leadership, management, and teaching positions.

This book is not intended to be a complete "how-to," step-by-step manual to the GPR method for archaeological mapping. Its goal, instead, is to introduce archaeologists to the method both theoretically and methodologically, with examples of both successes and failures. Complicated formulas, electronic wiring diagrams, and especially step-by-step instructions on how to work each GPR device by choice are *not* included. There are simply too many systems available, and a corresponding abundance of processing and image generating programs for each. To delve into the details of their own chosen GPR system or software package, readers will have to refer to the cited reference material, equipment manuals, or other technical sources, which are continually being modified and advanced by each GPR system manufacturer and researcher.

Most published articles that include any component of archeological geophysics always emphasize the successes, often with striking images of spectacular buried features, leading many to believe, often erroneously, that one or the other method is the greatest thing in archaeology since the invention of radiocarbon dating. This tendency to focus only on geophysical successes, while relegating failures elsewhere, is something all archaeological geophysicists are guilty of. Unfortunately, it leaves the impression on some that geophysics can do most anything, or on others that one method "works well" while others may not. These impressions are unfortunate, especially when an attempted survey cannot do what is desired for one reason or another, leaving the geophysically uninitiated with the erroneous impression that some techniques "don't work," giving all geophysical methods a bad name. It is therefore extremely important that geophysical archaeology be done in a deliberate manner, allowing for multiple working hypotheses to be tested and modified during the course of data collection and processing. Most important, when a survey is not successful in producing the results anticipated, the geophysical data and other information about the site should be reanalyzed and evaluated in order to determine why the final product did not produce the desired results. All the factors that could conceivably have affected the quality of the final product each step of the way must be understood by all involved in order for any results to be useful. Unfortunately, this type of thoughtful and reflective analysis is rarely the case, as most archaeologists neither want to nor have the background to understand the complexities of both the geophysical methods employed and the nature of the geological and archaeological complexities they may have confronted in the near surface. This is one of the many ongoing problems this book hopes, at least partially, to overcome.

It is doubtful that a totally uninitiated enthusiast of geophysical archaeology will pick up this book, read it, and be able to immediately go to the field and collect and interpret meaningful GPR data. It takes most beginners a fair amount of time to become proficient with the mechanics of data collection, transfer, processing, map making, and, most important, interpretation of results. Often it is necessary first to collect data, process them, perhaps test a site with auger holes or excavations for subsurface confirmation, and then think about what the maps that have been produced are illustrating about what lies below the surface. In all cases, GPR is measuring "something" in the ground, but determining exactly what that is always takes some thought and, most important, experience.

It is hoped, therefore, that this book will serve three purposes: first, to initiate archaeologists to GPR so they can begin to understand why and how the method works, depending on the problems to be solved; second, to provide a sufficient background so that archaeologists can set up their GPR surveys in a manner that the data collected will have the correct acquisition parameters for production of optimum images; third, once data are processed and interpretation becomes necessary, to provide a background with which to evaluate what is actually being measured and illustrated, and how those measurements change with differing field conditions. When GPR data are processed, but may not be immediately interpretable (a fairly common occurrence for the unexperienced), it is hoped that readers will be able to refer back to this book and other references in order to rethink what was done. They may need to reprocess databases or even repeat data collection or processing steps in a different way, sometimes with different tools, in order to refine and modify techniques for more positive outcomes.

Geophysical data collection and interpretation are not like learning to type or use a generic software program. Each site where geophysical data are collected is different in many ways, archaeologically, geologically, and environmentally. The variables that must be controlled for, and the collection and processing techniques that must be modified to produce a usable final product, are numerous, which is what makes the methods described here both challenging and potentially rewarding.

This book will begin with a discussion of GPR theory and method so that readers can become versed in how radar waves travel and are reflected in the ground. Only then will field setup and GPR collection procedures become meaningful, as they must be modified every time a new site is studied for what are always unique conditions. As simple as some field setup and calibrations seem to many, the thought that goes into a geophysical survey at the outset with regard to grid size, transect orientation and spacing, and ground surface preparation can often drastically affect a survey's final outcome. Often general equipment setup procedures must be altered and modified based on knowledge of both surface and subsurface ground conditions, size and orientation of potential archaeological features, and surface irregularities and disturbances. Most important, the type of the equipment being used and its unique acquisition parameters must be modified and possibly altered when information about the site conditions becomes apparent. If survey methods and procedures are not well thought out before the first bits of data are collected, the accuracy and interpretability of the final product will often be in doubt.

Appropriate processing and interpretation techniques for the GPR method are also an extremely important part of the technique and often the most difficult to master. Processing methods are constantly changing as new computer software is developed and improvements made to older "standard" techniques. As a result of this rapidly changing software environment, specific programs and techniques will not be discussed in detail, as this information would be made quickly obsolete. Therefore, only general methods, which are common to most GPR processing programs, will be covered. Each reader must search out the most appropriate programs for his or her needs, with the help of others working and developing methods for each GPR system.

Accurate and useful interpretation of the data, however, always comes with experience. Many successes and some possible pitfalls are included here, with information on how the data were processed and then interpreted for each. Readers, however, will likely be faced with new and different problems for each survey that is conducted, and therefore all techniques must by necessity remain somewhat fluid as different conditions are encountered and new and better field and laboratory methods are discovered and improved upon.

2
Introduction to Ground-Penetrating Radar

Ground-penetrating radar has a reputation as one of the more complex archaeological geophysical methods because it involves the collection of large amounts of reflection data from numerous transects within grids, often producing massive three-dimensional databases. The ability to detect many interfaces at different depths below the surface, the interpretation of those numerous reflections, and the difficulty in correlating them in many profiles within a grid therefore make GPR collection and processing a somewhat intimidating venture at first for the uninitiated. But its ability to produce high-quality three-dimensional images of the subsurface more than makes up for the method's relative complexity in data acquisition and processing.

Ground-penetrating radar data are usually collected along closely spaced transects within a grid, each of which consists of many thousands of radar waves that have been reflected from interfaces in the ground. It is an active method that transmits electromagnetic pulses from surface antennas into the ground, and then measures the time elapsed between when the pulses are sent and when they are received back at the surface. Radar travel times are measured in nanoseconds, which are billionths of a second. As the antennas are moved along the ground surface, individual reflections are recorded about every 2 to 10 centimeters along transects, using a variety of collection techniques. The form of the individual reflected waves (called a *waveform*) that are received from within the ground is then digitized into a refection *trace*, which is a series of waves reflected back to one surface location. When many traces are stacked next to each other sequentially, a two-dimensional vertical profile is produced along the transect that the antenna was moved (figure 2.1). Thousands of reflection traces in many profiles within a grid can then be analyzed to produce both two- and three-dimensional images of what lies below the surface.

Each of the reflected radar waves processed into profiles is recorded in elapsed time from their transmission to reception back at the surface (called *two-way travel time*). This time can be converted to approximate distance in the ground, giving each of the reflections precise depth information not available in other near-surface geophysical methods. The amplitudes of the reflected waves are particularly important because their variations are directly related to changes in the physical and chemical properties of different materials in the ground. When those amplitude differences are mapped spatially and with depth, accurate three-dimensional maps and images of buried archaeological features can be constructed.

Ground-penetrating radar surveys allow for a relatively wide aerial coverage in a short period of time, with excellent spatial resolution of buried archaeological features and their related stratigraphy. Often 50-by-50–meter grids of reflection

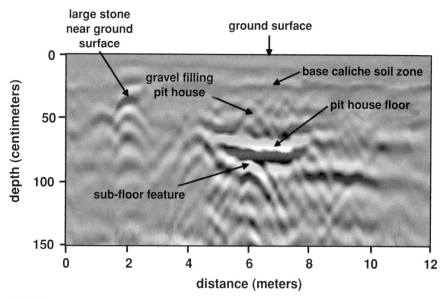

FIGURE 2.1
A GPR Reflection Profile. Distance along the profile is measured in meters, and two-way radar travel time, measured in nanoseconds, is converted to depth below the surface. This profile consists of 305 individual, sequentially stacked reflection traces. This profile was collected over a pit house floor near Alamagordo, New Mexico.

data can be collected in one day, with a 50-centimeter or less transect separation. Some radar systems have been able to resolve stratigraphy and other features at depths in excess of 40 meters, when soil and sediment conditions are suitable (Annan and Chua 1992; Bristow and Jol 2003; Davis and Annan 1992), but more typically, GPR is used to map features of archaeological interest at depths from a few tens of centimeters to 5 meters in depth. Radar surveys not only can identify buried features for possible future excavation but can also interpolate between excavations, projecting archaeological knowledge into areas that have not yet been or may never be excavated.

Ground-penetrating radar data are acquired by reflecting radar waves off subsurface features in a way that is similar to radar methods used to detect airplanes in the sky except energy is transmitted into the ground. Most GPR systems produce pulses from a surface antenna that are reflected off buried objects, features, or bedding contacts in the ground and detected back at the source by a receiving antenna that is located next to, or near, the transmitting antenna. As the radar pulses are transmitted through various materials on their way to the buried target features, their velocity will change, depending on the physical and chemical properties of the material through which they are traveling. When the travel times of the energy pulses are measured, and their velocity through the ground is known, distance (or depth in the ground) can be accurately measured.

Ground-penetrating radar units have recently become very portable, and complete systems can potentially be transported in one's backpack into remote areas (figure 2.2). Most systems are powered from any high-amperage battery such as a 12-volt car battery or with portable electrical generators or directly from 110- or 220-volt alternating currents. Some recently developed GPR units can power all of the radar equipment and computers necessary for data acquisition and field processing for many hours on a few small rechargeable batteries.

Some of the earliest model GPR systems recorded raw subsurface reflection data on paper printouts that precluded postacquisition processing. Although these radar systems, a few of which are still in use, can many times yield valuable subsurface information, the modern digital systems that record reflection data on a computer hard drive for later filtering, processing, and sophisticated data analysis are far superior. Most important, when the recorded data are in digital form, a computer can process, filter, and enhance the raw field reflections almost immediately after they are collected.

Accompanied by a trend in equipment miniaturization, rapid computer processing of acquired GPR data can now occur immediately after they are acquired,

FIGURE 2.2
GPR Equipment. A GSSI subsurface interface radar (SIR) system with a 400-MHz antenna, using the packing case as a field table.

and interpretation can often begin while still in the field, allowing archaeologists to produce three-dimensional images of buried features just minutes or hours after data are acquired. When this is done, additional GPR data acquisition or the planning of excavations to confirm discovered features of interest can also begin almost immediately, making geophysical data collection, interpretation, and excavation an often iterative process.

The success of GPR surveys in archaeology is to a great extent dependent on soil and sediment mineralogy, clay content, ground moisture, depth of burial, surface topography, and vegetation. It is not a geophysical method that can be immediately applied to all geographic or archaeological settings, although with thoughtful modifications in acquisition and data processing methodology, GPR can be adapted to a great variety of site conditions. In the past, it was assumed that GPR surveys would only be successful in areas where soils and underlying sediment are extremely dry and therefore nonconductive (Vickers and Dolphin 1975). Although radar wave penetration, and the ability to reflect energy back to the surface, is often enhanced in dry ground, good GPR data have been collected in areas that are wet and contain an abundance of clay (Conyers 2004). Conditions such as this had always been avoided in the past as they were considered poor GPR areas, but this is not always the case, as will be detailed in some of the examples in this book. In contrast, poor GPR data have been collected in very dry and sandy areas, which are usually considered good GPR areas, suggesting that many more factors affect radar transmission and reflection than have been documented in the geophysical literature.

Prior to conducting a GPR survey, it is important to take into consideration what types of equipment to use, the field collection methods that will be employed, and numerous data acquisition parameters. These factors will vary considerably depending on the geographic and geologic setting of the surveys, surface obstacles, ground conditions, and the depth of the archaeological features and stratigraphy to be studied. Many archaeological geophysical surveys are conducted precisely because little is known about what lies below the ground. When this is the case, it is difficult to take all these variables into consideration, especially the nature of soils and sediments and the depth and types of archaeological materials that might be encountered. Acquisition parameters and equipment must therefore often be adjusted and modified while in the field, once some preliminary data are collected and analyzed. When this is the case, preliminary calibrations that are necessary in order for optimal data acquisition can often be a somewhat nerve-wracking experience, especially if bystanders are looking on expecting immediate results and asking the most common and annoying question: "Have you found anything yet?" When this occurs, all one can do is politely explain that a great deal of thought must go into preliminary analysis of field conditions before any results are available, and then get back to work. As will be discussed later, a good deal of deliberation must be given to what will be the optimum GPR data collection procedures for each study area before the first useful reflection profile is collected and stored on the computer storage medium.

Once GPR data have been acquired in the field and recorded digitally on a computer, a wide variety of data-processing and interpretation techniques are available. Depending on the archaeological questions to be asked and the quality of the radar reflection data acquired, these processing techniques can also be varied and modified to fulfill specific needs. To be able to fully understand and interpret GPR reflection data, the user must first understand the basic theoretical aspects of the electromagnetic energy propagation as well as GPR collection methods. Only then will postacquisition data-processing techniques and interpretation of the final product become meaningful.

Some of the very detailed and complicated aspects of the GPR method, which are not immediately applicable to general archaeological investigations—such as complex electromagnetic theory, equations used in data processing, and details about the schematic components of the equipment—are held to a bare minimum in this book. For the average archaeological user, these somewhat esoteric subjects can be addressed by a radar technician or electrical engineer, or studied in greater detail from the cited references. This book includes only aspects of those subjects that are most important for collecting, processing, and interpreting most GPR databases for most archaeological applications.

Ground-penetrating radar acquisition is becoming simpler and data processing more intuitive for most archaeologists. Even so, as recently as the mid-1990s, some supposedly technically sophisticated archaeologists have complained that GPR data are too complicated, expensive, and difficult to process for the archaeological community (Meats 1996). Many archaeologists, especially in Europe, still prefer magnetic and resistivity methods, because they have a longer history of success there and data from those surveys are usually less complicated to acquire and process. In the last few years, however, GPR has moved into the archaeological mainstream worldwide and is no longer a method reserved only for geophysicists with "black boxes" who perform some kind of "magic" in the field. Most archaeologists trained today have more than enough scientific background and computer skills to allow them to understand and use this high definition three-dimensional method. All it usually takes is some field experience, the background that will allow prudent acquisition procedures, a determination to try it out, and the patience to process and interpret data once they are acquired.

HISTORY OF GPR IN ARCHAEOLOGY

The first large-scale application of radar was during World War II when the British and later Americans used crude but effective systems to detect airplanes. The word *radar* is actually an acronym that was first coined in the 1930s for *ra*dio *de*tection *a*nd *r*anging (Buderi 1996). The first attempt at what could be called ground-penetrating radar was made in Austria in the 1920s to determine the thickness of ice in a glacier (Stern 1929). The ground-penetrating aspects of radar technology were then largely forgotten until the late 1950s when U.S. Air Force radar technicians on board airplanes noticed that their radar pulses were penetrating the glacial ice when flying over Greenland. A number of mishaps occurred because airborne radar analysts were detecting the bedrock surface below the ice and interpreted it as the ground surface, neglecting to see the large thickness of ice above and leading planes to crash into the glaciers. This realization that radar would readily penetrate ice ultimately led to numerous investigations into the ability of radar to detect a number of subsurface interfaces, including soil properties and the groundwater table. In 1967, a prototype GPR system was built by the National Aeronautics and Space Administration (NASA) and sent on a mission to the moon in an attempt to determine surface conditions prior to landing a manned vehicle (Simmons et al. 1972).

The applicability of GPR to locate buried objects or cavities such as pipes, tunnels, and mine shafts (Fullagar and Livleybrooks 1994) was immediately recognized in the 1970s, and its widespread use as a geotechnical tool began. Methods were soon developed to define lithologic contacts (Baker 1991; Basson et al. 1994; Bristow and Jol 2003; Jol and Smith 1992; van Heteren et al. 1994), faults (Deng et al. 1994), and bedding planes and joint systems in rocks (Bjelm 1980; Cook 1973, 1975; Dolphin et al. 1974; Moffatt and Puskar 1976). Soil scientists and hydrologists also began using GPR to investigate buried and surface soil units (Collins 1992; Doolittle 1982; Doolittle and Asmussen 1992; Doolittle and Collins 1995; Freeland et al. 1998; Johnson et al. 1980; Olson and Doolittle 1985; Shih and Doolittle 1984) and the depth and nature of the groundwater table (Beres and Haeni 1991; Doolittle and Asmussen 1992; van Overmeeren 1994, 1998). More recent work has shown the utility of GPR for mapping specific packets of sediment for the definition of ancient depositional environments (Bridge et al. 1995; Bristow et al. 1996; Bristow and Jol 2003; Jol et al. 1996; McGeary et al. 1998; van Overmeeren 1998). The applicability of using GPR techniques for locating unexploded ordinance and land mines has also been studied, with great promise (Bruschini et al. 1998; Daniels 2004). Civil and structural engineers have used GPR to map road pavement structures and have applied those data to the inspection of the interior of many different media (Hugenschmidt et al. 1998). Forensic scientists and law enforcement agencies' desire to find buried bodies or other materials has expanded the use of GPR in a number of instances, locating graves and sometimes actual human remains of murder victims or other bodies in the ground (Davenport 2001a, 2001b; Davis et al. 2000; Ivashov et al. 1998; Nobes 1999; Strongman 1992).

The archaeological community was quick to grasp the potential of using GPR to locate and help define buried archaeological features and associated sediment and soil layers. One of the first applications to archaeology was conducted at Chaco Canyon, New Mexico (Vickers et al. 1976), in an attempt to locate buried walls at depths of up to 1 meter. A number of experimental antenna traverses were made at four different sites and paper reflection profiles were analyzed in the field. It was determined that a few of the anomalous radar reflections represented the location of buried walls.

These rudimentary studies at Chaco Canyon were followed by a number of GPR applications in historical archaeology. Radar surveys were successfully used in the search for buried barn walls, stone walls, and underground storage cellars (Bevan and Kenyon 1975; Kenyon 1977). In these early studies what were described as "radar echoes" were recognized as being generated from the tops of buried walls, and depth estimates were made, using approximate velocity measurements for local soil characteristics.

Initial successes in historical archaeological applications were followed in the late 1970s at the Hala Sultan Tekke site in Cyprus (Fischer et al. 1980) and the Ceren site in El Salvador (Sheets et al. 1985). Both of these GPR surveys produced unprocessed reflection profiles in the form of paper records that were successful in delineating deeply buried walls, house platforms, and other buried archaeological features. These initial successes were primarily a function of very dry electrically resistive matrix material that was relatively "transparent" to radar energy propagation, allowing for deep energy penetration and producing relatively uncomplicated reflection records from buried archaeological features that were easy to interpret.

During 1982 and 1983, GPR surveys were conducted at a historic site in Red Bay, Labrador, in an attempt to locate graves, buried artifacts, and house walls associated with a sixteenth-century Basque whaling village (Vaughan 1986). This area was an extremely challenging test for archaeological GPR mapping because the soils were wet, and the overburden contained large cobbles and complicated stratigraphy that produced a variety of difficult to interpret reflection records. Nonetheless, artifacts and archaeological features that were buried by up to 2 meters of beach deposits and peat were discovered in many of the GPR profiles, which were later excavated and confirmed. This study is notable because velocity tests were performed and radar travel times to potential archaeological targets were corrected to approximate depths in the ground prior to being uncovered. It was determined that grave goods, consisting of bone, did not contrast enough with the surrounding beach deposits to appear as distinct reflections, but disturbed soil in some graves appeared as anomalous reflection zones on some radar profiles. Other significant reflections were found to have been generated from buried walls that consisted of piles of beach cobbles used for walls and foundations, but they were difficult to discriminate from other random rocks.

A comprehensive series of GPR surveys were conducted in Japan in the mid-1980s in order to locate buried sixth-century houses, burial mounds, and what were termed "cultural layers" (Imai et al. 1987). These studies were successful in identifying ancient pit dwellings with clay floors, which were buried in some cases by as much as 2 meters of volcanic pumice and loamy soil. The interface of the house floors with the overlying pumice produced very distinctive reflections that were readily recognizable on GPR profiles. Much of the site discovered by GPR mapping was then excavated to confirm the results. Three distinct stratigraphic horizons (their cultural layers) were found to be buried soil horizons containing many stone artifacts that were discarded during different periods of occupation. This important conclusion allowed for the mapping of these distinct soils and the associated archaeological features on and buried within them throughout portions of the site that had not been excavated.

Throughout the late 1980s and early 1990s, GPR continued to be used successfully in a number of archaeological contexts, but in most cases, these studies were what could be called "anomaly-hunting" exercises. Usually unprocessed or partially processed GPR profiles were viewed as paper records, or on a computer screen as they were acquired, and interesting reflections, which could possibly have archaeological meaning, were excavated. Unfortunately, the inability to discriminate archaeological from geological reflections in these studies often left archaeologists with the impression that GPR was a "hit-or-miss" method at best.

Prior to 1993, the most encompassing and successful archaeological application of GPR was that employed in the mapping of the houses and burial mounds in Japan, already discussed (Imai et al. 1987). These successes were followed up by numerous additional GPR surveys in Japan, conducted by Dean Goodman and his colleagues (Goodman and Nishimura 1993; Goodman 1994, 1996; Goodman et al. 1995; Goodman et al. 1998). Great strides were made in these studies that benefited from advances in computer processing speed and the development of software programs written by Goodman specifically for archaeological GPR data processing.

In the early 1990s, GPR manufacturers began to market systems that could collect reflection data as digital files, storing large amounts of data for later processing and analysis. About this same time, inexpensive and increasingly powerful personal computers were also becoming available that could process these digital data in ways that were not previously possible, at least on the typically low archaeological budgets. The pioneering studies of Goodman and his collaborators led to many important GPR acquisition and data-processing techniques, including amplitude slice maps, computer-simulated two-dimensional models, and three-dimensional reconstructions of buried features (Conyers and Goodman 1997; Goodman 1996; Goodman et al. 1998). The Japanese GPR studies discovered and mapped a wide variety of buried archaeological sites, including ceramic kilns, burial mounds surrounded by moats, and individual stone-lined burials (Goodman et al. 1995). In this work, a wide range of burial conditions were encountered, which were computer modeled prior to data acquisition in order to determine the best equipment to use and the setup configurations that would likely work best.

The realization that radar reflections, measured in time, could be defined in real depth when radar wave velocity was determined was one of the major GPR advancements for archaeology (Vaughan 1986; Imai et al. 1987). The identification of reflections that correspond to horizons of archaeological interest was also used in a limited way to map related stratigraphy and buried topography (Imai et al. 1987; Conyers 1995; Conyers and Spetzler 2002). Recently, the application of two-dimensional computer simulation and three-dimensional processing techniques (Conyers et al. 2002; Goodman et al. 1995, 1998, 2004) has shown that even radar data that does not yield immediately visible reflections can still contain valuable reflection data when computer processed. The use of these and other new processing and imaging techniques has recently greatly expanded the utility and speed of GPR exploration and mapping in archaeology.

Many archaeologists who employ GPR at their sites are mainly concerned only with identifying buried anomalies that represent features of interest (see, e.g., Butler et al. 1994; Sternberg and McGill 1995; Tyson 1994; Valdes and Kaplan 2000). Although this type of GPR application is valuable, in that buried features can be immediate identified, this book will illustrate how the radar reflection data acquired in these types of studies can be further enhanced by a number of computer processing, interpretation, and display techniques. With little additional effort, recently developed computer technology allows for the construction of maps that can be interpreted in ways that will yield much more information about a site than was thought possible just a few years ago.

In the future, GPR's ability not only to noninvasively map buried structures and other cultural features in real depth, but also to reconstruct the ancient landscape of a site and human interaction with it, will become increasingly important. Computer filtering and enhancement of GPR reflection data is also becoming widespread as researchers increase their familiarity with some of the computer-processing techniques discussed in this book and many others that are presently available. Recent research has demonstrated and quantitatively accessed the differences in data quality that varies with antenna frequencies, the spacing of transects and the density of reflections along transects, and the types of data analyses used to display the final product (Neubauer et al. 2002).

The ability of GPR to collect data in a three-dimensional block has recently led some researchers to begin analyzing the reflected wave amplitudes in complex and exciting ways (Conyers et al. 2002; Goodman et al. 1998, 2004; Leckebusch and Peikert 2001; Leckebusch 2003; Moran et al. 1998). If the higher amplitudes can be shown to denote the location of important buried archaeological features, then their locations in three dimensions can be visualized using a number of imaging software programs. In this way, the lower-amplitude reflections are effectively removed from the data set, leaving only those of importance, which can be visualized three-dimensionally. The location of certain radar amplitudes in space, which are proxies for the actual location of features in the ground, are then rendered to produce "virtual reality" images of what lies below the surface. This has been done by cutting through the block (Neubauer et al. 2002) or rendering out only the higher amplitudes and presenting the final product in three-dimensional, rotating images or videos (Conyers et al. 2002; Goodman 1998; Goodman et al. 2004; Heinz and Aigner 2003; Leckebusch and Peikert, 2001; Piro et al. 2003).

One area of study that has been given much recent attention is the integration of multiple data sets collected from different geophysical methods. This integration has been accomplished by fairly simple mathematical correlation (Savvaidis et al. 1999), numerical comparison (Piro et al. 2000) of GPR amplitudes in slice maps with other geophysical maps, the use of statistical modeling (Marukawa and Kamei 1999), and within geographic information systems programs (Kvamme 2003). Experiments have also been progressing in which multiple antennas are used to collect reflections simultaneously to enhance three-dimensional subsurface resolution (Pipan et al. 1996).

3
GPR Theory and Practice

RADAR ENERGY GENERATION AND PROPAGATION

Radar waves that move in both the air and the ground are a form of electromagnetic energy composed of cojoined oscillating electrical and magnetic fields (figure 3.1). These waves are produced when an electric current oscillates back and forth in a conductive body, producing a subsidiary magnetic field (Kraus 1950; Rojansky 1979). Electromagnetic waves are then generated that propagate outward from the source, with the electrical portion of the waveform moving perpendicular to the magnetic. If either the magnetic or electrical component of the field is lost (attenuated, absorbed, or conducted away), the wave will cease propagating and die. Propagation of radar waves occurs readily in air or space, and unless they encounter a medium that absorbs or reflects them, they will travel an infinite distance. Radar waves are capable of penetrating up to a few meters or more in some ground conditions before they are attenuated and the energy is lost.

Radar energy used in GPR is produced at an antenna, the simplest of which is a copper wire or plate on which an oscillating electrical current is applied. Depending on the frequency of the oscillation (measured in cycles per second), different wavelengths of propagating radar waves are produced. The higher the oscillation frequency, the shorter the wavelength of electromagnetic energy produced, and vice versa. To generate long radar wavelengths, larger antennas with a lower oscillation frequency are necessary. Each wavelength of propagating energy will behave differently in different media within the ground, with longer wavelength energy usually propagating deeper with less reflection from small objects and shorter wavelengths

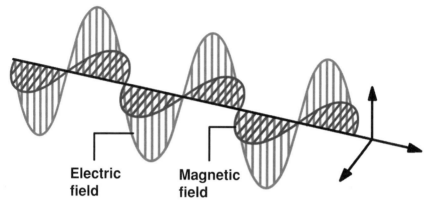

Electric field **Magnetic field**

FIGURE 3.1
Electromagnetic Wave Propagation. An electromagnetic wave consists of cojoined electrical and magnetic waves that feed on themselves during propagation.

penetrating to only shallow depths, but reflecting much more readily from smaller buried discontinuities. The shorter wavelengths penetrate less deeply because they are more readily attenuated by most ground conditions (Leckebusch 2003).

There are many different designations assigned to electromagnetic waves, each defined by its wavelength (which is determined by the frequency of the oscillating source that produces them). Visual light is the most commonly recognizable electromagnetic wave, as are X rays, ultraviolet and infrared radiation, TV, radio and cellular phone transmissions, gamma rays, and many more. Radar waves used by most GPR systems occupy a specific portion of the radio spectrum (figure 3.2).

Frequency of propagating radar waves is measured in units of hertz, which is defined as cycles per second. Gamma rays, X–rays, and visual light have very high frequencies of oscillation on the order of 10^{12} to 10^{17} cycles per second. These very high frequencies produce extremely short wavelength energy, measured in fractions of millimeters. Radio waves, a subset of which are radar waves, have much lower frequencies, with wavelengths of propagating energy that vary from a few centimeters to at most a few tens of meters in length. The radar energy used in most GPR applications has frequencies ranging between about 10 and 1,500 megahertz (figure 3.2). This energy occupies a portion of the same electromagnetic spectrum as television and FM radio, cellular phones, and other personal communications devices (figure 3.2), which has recently been the cause of some (mostly irrational) concern by government regulators who are worried about possible GPR antenna interference with communication transmissions (Chignell 2004).

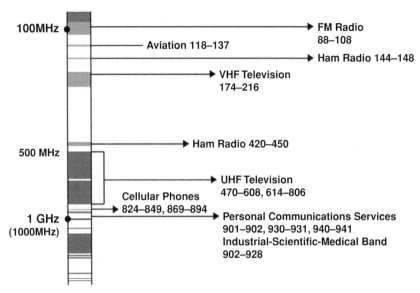

FIGURE 3.2
The GPR Frequency Distribution. Most GPR antennas operate within the frequency band used by many communication devices. There can be a good deal of interference of these usages with some GPR antennas, especially in the 500- to 1,000-MHz frequencies that often overlap television, cellular phone, and pager transmissions.

The GPR method is based on the transmission of electromagnetic pulses, which then propagate as waves, into the ground and measuring the time elapsed between their transmission, reflection off buried discontinuities, and reception back at a surface radar antenna. Each physical or chemical change in the ground through which the radar waves pass will cause some of that energy to be reflected back to the surface, while the remainder continues to propagate deeper until it finally dissipates. Buried discontinuities where reflections occur are usually created by changes in the electrical or magnetic properties of the rock, sediment or soil, variations in their water content, lithologic changes, or changes in bulk density at stratigraphic interfaces (VanDam and Schlager 2000). Reflections also are generated when radar energy passes across interfaces between archaeological features and the surrounding matrix. Void spaces in the ground, which may be encountered in burials, tombs, tunnels, caches, or pipes, will also generate significant radar reflections because of a similar change in radar wave propagation velocity. Many bed boundaries and other discontinuities in the ground will reflect a *wavelet* of energy (a positive and negative amplitude wave) back to the surface to be recorded (figure 3.3). A composite of

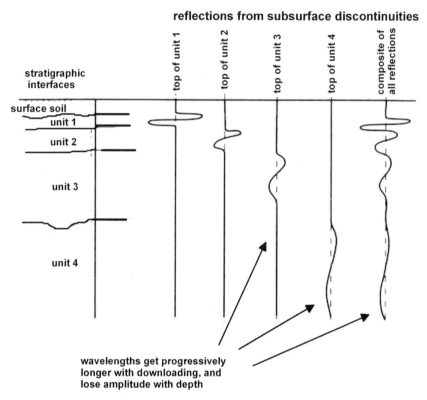

FIGURE 3.3
The Generation of a Waveform. A waveform is a composite reflection trace produced by recording a number of wavelets generated at many subsurface interfaces in the ground.

many wavelets that are recorded from many depths in the ground produces a series of reflections generated at one location, called a *reflection trace* (figure 3.3).

GENERATION AND RECORDING OF GPR WAVES

To collect GPR reflections, paired antennas are moved along the ground in transects (figure 3.4). One antenna generates propagating radar waves, and a second paired antenna records the reflection traces generated from reflections below. When many hundreds or even thousands of reflection traces are stacked sequentially as they are collected along an antenna transect, a reflection profile is produced, as in figure 2.1. With the aid of a computer, reflections from thousands of traces within many profiles in a grid can be converted to depth in the ground,

FIGURE 3.4
GPR Antennas Collecting Data. Reflection profiles are collected by moving antennas in transects. This fiberglass box houses paired transmitting and receiving 400 MHz antennas. Energy is transferred to and from the control system by means of a cable.

and the strength of the reflected waves can be analyzed, producing images that portray the nature of materials in the ground in three dimensions.

Radar antennas are usually housed in a fiberglass or plastic sled that is placed directly on the ground (figure 3.4) or supported on wheels a few centimeters above the ground. Commercial GPR systems can have antennas that are mounted in a number of different ways, but all attempt to place them at or near the ground surface. When two antennas are employed, which is almost always the case, one is used as a transmitter and the other as a receiving antenna. This is called *bistatic mode*. A single antenna can also be used as both a transmitter and receiver, in what is called a *monostatic* system. In this type of data collection, the same antenna is turned on to transmit a radar pulse and then immediately switched to receiving mode in order to measure the returning reflected energy received from within the ground. There are some multichannel GPR systems commercially available that can send and receive from multiple antennas simultaneously, but so far they have not been commonly applied to archaeology because of their size, expense, and the complexity of data processing. Researchers are also developing antenna arrays that can potentially receive reflections at tens or even hundreds of surface antennas to produce accurate three-dimensional images of the subsurface, but these also have seen limited archaeological application (Leckebusch 2000).

Antennas are usually hand-towed along survey transects (figure 3.4) at about the speed that someone can walk, but they can also be pulled behind a vehicle, towed by a boat on a lake (Leckebusch 2003), or even suspended from a helicopter. Most GPR systems are capable of generating and collecting reflection traces at a very high rate and can be pulled behind vehicles on a roadway at quite high speeds, but this collection mode is usually not practical or desirable for most archaeological applications.

Ground-penetrating radar systems also have the ability to collect individual reflection traces in steps along a transect instead of being moved continuously. During step acquisition, the smaller the spacing between steps, the greater the number of reflection traces recorded per unit distance, with a corresponding increase in subsurface coverage and therefore resolution. The step acquisition method, however, necessitates more field time because the antennas must be manually moved to each step for each reflection trace to be recorded and therefore less data can be acquired in a given amount of time. Most systems can be programmed to collect data with a survey wheel, or some similar device that can measure where the antennas are in distance along each transect, which can expedite data processing as all recorded reflection traces can be assigned a specific surface location (figure 3.5). A number of prototype systems that are in the experimental stage use global positioning systems or self-tracking laser theodolites to measure distance and location of the antennas on the ground, but they have not as yet been commonly used in archaeology (Lehmann and Green 1999).

ACQUISITION PROCEDURES

Usually antennas are placed directly on the ground surface or as close to the ground as possible. If antennas are located too far above the ground surface, energy will not effectively couple with and then penetrate into it. When this happens, much of the transmitted energy is reflected back to the receiving antenna from the air–ground interface, leaving little to penetrate more deeply.

Although the source of the transmitted energy can be thought of as one distinct radar pulse generated from the surface antenna, this perception is not technically correct. Most GPR systems transmit radar pulses at extremely high rates ranging from 25,000 to 50,000 pulses per second, and digitizers in most systems are not fast enough to sample reflected waves received from any one of these distinct pulses (Leckebusch 2003). To overcome this problem, radar control systems use incremental sampling methods that produce a composite reflection trace by recording the first digital sample within a trace from reflected energy that arrives from the first transmitted pulse. The second sample is recorded from the second pulse and so on until one complete reflection trace is constructed. It may, therefore, take 512 or more samples, from the

FIGURE 3.5
An Antenna Survey Wheel. Antennas can be attached to survey wheels, which can be programmed to collect a given number of reflection traces every programmed distance along a transect.

same number of consecutively produced pulses, to compile the record of one complete reflection trace (figure 3.6). Most GPR users collect a minimum of 512 digital samples to record each reflection trace, although more or less can be defined in most system setup commands.

In continuous data acquisition, if the antennas are moving across the ground at an average walking speed, incremental sampling will create some averaging of the recorded reflections as conditions change in the subsurface. This averaging procedure is usually negligible because it is occurring very quickly and would affect the recorded data only if the subsurface features and layers were extremely variable or the antennas were moving at a high speed along the ground.

The step method of data acquisition uses the same general method, except discrete reflection data in one reflection trace are received and recorded at every step interval. When the antennas are moved to the next step, a reflection trace is again acquired and recorded digitally as 512 or more samples. If the antennas are set up for acquisition in the step method, a beeper or some other form of notice

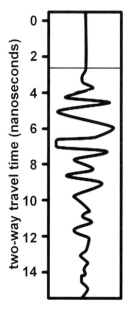

FIGURE 3.6
A Reflection Trace. This reflection trace shows the ground surface at about 2.5 nanoseconds, with the waveform losing amplitude with time, as energy is attenuated in the ground; 512 digital samples are recorded to define this one reflection trace.

is given after each reflection trace is acquired at a station, telling the antenna handler to move the antennas to the next station along a presurveyed transect.

The movable radar antennas are often connected to the GPR control unit by cable (figure 3.4). Some systems record the reflection data digitally directly at the antenna, and the digital signal is sent through fiber optic cables back to the control module (Davis and Annan 1992). Others send an analog signal from the antennas, through coaxial cables, to the control unit where the reflection waveforms are digitized and stored. In the near future, wireless transmission of data from the antennas to a base station will also be possible, allowing all cables to be dispensed with. Many manufacturers are marketing systems that can be used by one person, with the GPR control unit, power sources, and antennas all placed on a wheeled cart or carried on a backpack for nontethered transport within a grid (figure 3.7).

To create a vertical display of the subsurface reflections, all recorded reflection traces, no matter what the acquisition method, are displayed in a format where the two-way travel time or approximate depth of the reflected waves is plotted

FIGURE 3.7
A One-Person GPR System. Some GPR systems can be mounted on a cart for one-person operation.

on the vertical axis with the surface location on the horizontal axis (figure 2.1). In standard two-dimensional reflection profiles that are produced by moving the antennas continuously across the ground, pulses of radar energy are generated at a set time interval, and the horizontal scale will vary because of changes in the speed at which the antennas are moved. Depending on variability in the speed at which the antennas are pulled along the ground, the number of reflection traces collected per unit of distance covered will also vary, making the horizontal scale nonlinear. When data are collected in this fashion, manual markers are placed in the data string at known distances along the transect, and the horizontal scale of the profiles can be adjusted later (Shih and Doolittle 1984). These location markers placed in the recorded reflection data are called *fiducial marks*. Computer processing methods (discussed in chapter 6) can correct the horizontal scale, interpolating between the fiducial marks and either expanding or contracting the spacing of reflection traces between these known locations to create a linear evenly spaced horizontal scale. In the step acquisition or survey wheel method, the horizontal scale is set by the distance between acquisition stations, or the number of traces programmed to be collected per unit distance along the ground, and no horizontal adjustment is usually necessary during postacquisition processing.

The vertical scale in all GPR profiles is measured in two-way travel time, but it can be converted to approximate depth if the velocity of radar energy in the ground is known. Postacquisition computer processing, discussed in chapter 6, typically adjusts both vertical and horizontal scales to create profiles with any desired vertical or horizontal scale or exaggeration. If there is significant elevation variation along a survey transect, topographic corrections should also be made that will adjust the recorded reflections for surface irregularities (figure 3.8). This can only be accomplished if a ground surface topographic survey is made.

Three people are the optimum number of workers usually necessary to conduct a GPR survey, although with newer, more portable systems, one person can sometimes do it alone, with lots of bending over to move and straighten survey tapes as antenna transects are moved within a grid. If more than one person is available, one helper is usually assigned to pull the antennas, or move them in steps, along the surveyed grid transects. Another can stay at the control unit to view the reflection data coming across the computer screen and make notes of significant reflections as they are recorded. If neces-

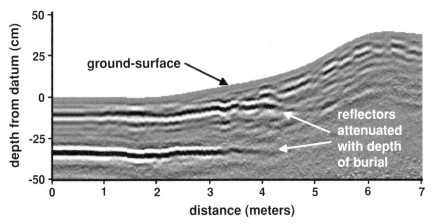

FIGURE 3.8
Topographically Corrected Reflection Profile. This is a reflection profile where radar travel times have been converted to depth and then adjusted for a sloping ground surface. The energy that produced two high-amplitude reflections in this profile is greatly attenuated with progressive depth of burial.

sary, the person at the control unit can also activate a surface location marker button by a command from the person pulling the sled to note the location of obstacles or distance markers along antenna transects. A third person comes in handy to make sure antenna cables do not snag on surface obstacles, help maneuver the antennas when necessary, clear surface debris away from transects, and move tape measures or survey lines for each transect. Some of this labor can be saved if a cart or backpack is used to move the control system with the antennas.

The most tedious, but also important, part of a GPR survey is performed by the person moving the antennas. This job is most difficult during continuous data acquisition because the person pulling the sled must not only walk backward but must also make sure that the antennas are moving parallel to the designated transect line marked by a tape measure or survey rope. If data are being acquired in continuous acquisition mode, where radar pulses are being generated at a programmed number per second, the antenna puller must also pay attention to when the antennas move past designated surface markers. At each presurveyed location, a marker button on the antenna handle must be pushed to place fiducial marks in the reflection records. When a survey wheel is used or antennas are moved in steps, manual marks of this sort are not necessary, and antenna pulling is an easier task. Another important aspect of moving the antennas along the ground is making sure that the antennas are in the same orientation and the same distance above the ground or are directly touching it at all times. Changes in antenna orientation with respect to the ground can potentially cause variations in the recorded reflections that can be confused with "real" material variations in the ground. This phenomenon is called *coupling loss*, and its ramifications will be discussed later in this chapter.

Prior to acquiring reflection data along a transect, the person working the control unit needs to note in a field book the transect number in the grid and the corresponding file number to which the reflection data will be saved. This task cannot be understated because even the most sophisticated computer storage systems can sometimes mysteriously "lose" data (usually because of operator error), and only a handwritten record will reconstruct the procedures used in the field. Surface obstructions that might be encountered, and large trees or other features both on the surface and in the ground that could possibly reflect radar energy, also need to be documented for each transect and their approximate location noted. Many other acquisition procedures more common to geologic studies are discussed by Jol and Bristow (2003).

COLLECTION OF TRANSECTS IN GRIDS

Most archaeological applications necessitate the acquisition of GPR data in a rectilinear grid over the area to be surveyed (Doolittle and Miller 1991). If the grid is oriented to the north, then survey transects can be acquired in either north-south or east-west orientations (or both, if desired). The grid should be situated so that surface obstacles are avoided, and it should be located on the most even and horizontal ground possible. Unless data are being collected on a paved parking lot or mowed field, this is rarely the case. When surface obstructions are present, a grid pattern with survey transects of different lengths can also be easily constructed to avoid the obstacles, creating a complex grid pattern. Computer programs are available that will process reflection data from any grid pattern, as long as the antenna transects are parallel or perpendicular to each other. If buried pipes, tunnels, or electrical cables are known to be present, their location should also be identified in advance, and the grid should be located so as to avoid them. When this is not possible, their location can often be determined by identifying reflections derived from them when the data are interpreted.

Rectangular or rectilinear grids are preferable to other grid designs for a number of important reasons. Digital reflection data collected as parallel or perpendicular transects in a grid can easily be exported to computer display and imaging processing programs that are almost always preset for this gridding method. In this way, the data can be quickly processed and interpreted without time-consuming transect surveying and spatial rectification. In addition, with a rectangular grid, important reflections in each profile can be immediately correlated to others, and reflections can be visually "tied" to parallel or perpendicular transects throughout the grid. In all cases, a sketch of the grid with notes on the transect length, orientation, and beginning and end locations of each transect should be made.

Simple rectilinear grid patterns are not always possible or desirable when conducting some GPR surveys. In some cases, surface conditions or time constraints may necessitate a series of separate nonparallel transects that can still yield good subsurface coverage. Care must be taken when setting up a nonrectangular grid so that the locations of all transects are accurately surveyed and the acquired reflection data can be accurately mapped in three dimensions. When grids of this sort are collected, reflection data must usually be manually interpreted, and map making can be both time-consuming and tedious. Important results can still be obtained, however. Surveys transects that radiate outward from one central area have been sometimes used—for instance, to define a moat around a central fort-like structure (Bevan 1977). A rhomboid grid pattern has also been used with success in Central America within a sugarcane field on the side of a hill where antennas, by necessity, had to be pulled between planted cane rows (Conyers 1995). Sinuous reflection profiles, none of which were either straight or parallel to each other, were collected as antennas were towed over the surface of a lake in Switzerland (Leckebusch 2003). A very complicated grid of sinuous profiles was the product of the boat drifting away from floating survey lines, which necessitated some complex spatial adjustment before interpretation. Although all these non-rectilinear surveys necessitated additional data-processing time to spatially rectify all recorded reflections, they ultimately proved just as useful as those acquired in rectangular grids.

All grids must be accurately surveyed and placed within an overall site map using some type of surveying technique either before or after the acquisition of the GPR data. At the very minimum, the corners of each grid (if rectangular) must be accurately located, and care must be taken so that all transects are parallel or perpendicular to each other within it. This part of the GPR acquisition process can sometimes be the most time-consuming and tiresome part of a survey, but it is extremely important. In the near future, it will be possible to survey grids in a more cursory fashion and use a global positioning system (GPS) to automatically record the location of survey transects (Czarnowski et al. 1996). In this way, the exact coordinates of each transect, and the elevation of the ground surface along them, will be automatically recorded as digital data on a separate channel during acquisition. This technology, which is becoming more common in geological data acquisition, is just beginning to be applied to GPR systems, and at present, traditional survey methods conducted with a transit and rod, laser theodolite, or tape and compass must still be used.

When the ground surface is rough, uneven, or sloping, closely spaced topographic elevations along each survey transect must be obtained so that corrections of subsurface reflections can be made during postacquisition processing (Sun and Young 1995). If the ground is evenly sloping, it may only be necessary to survey the beginnings, ends, and a few elevations along each transect, or at each change of slope, and then interpolate elevations in between to save surveying time. When surface irregularities are numerous, elevation surveying must be done at more frequent intervals (perhaps every meter or less), and data processing becomes more of a chore. The location of any surface feature that could conceivably reflect radar energy should also be mapped at the same time. Trees, overhead branches, houses, or other objects must be accurately located and placed on survey maps so that when the reflection data are later processed, reflections that might have been generated from them can be factored out.

Occasionally there is a need to immediately determine the location of important subsurface reflections with no real need to produce a regional map of the site. If this were the case, raw reflection profiles could be collected and interpreted immediately as the antennas are randomly pulled across the ground. It may not even be necessary to set up a grid, and one could conceivably just wander around a site, producing reflection records until a reflection anomaly is visible on the computer screen. This method is commonly used for buried utility detection at construction sites. Significant reflection locations could then be marked on the ground with pin flags or chalk as they are identified on the computer screen. This is an extremely easy way to conduct a GPR survey (and quite fun, as one gets instant results) but it is full of pitfalls in most archaeological contexts. For the most part, it is difficult to immediately identify important reflection anomalies in raw reflection profiles as they are appearing on the computer screen or on paper printouts. Often important reflections do not appear (or are not recognizable) until the antennas have moved past the subsurface feature producing them, and then one must estimate their location after the fact. This "instant results" type of data acquisition and reflection profile interpretation method should never be used in place of the more standard data acquisition method, which is to collect straight profiles in rectilinear grids for later processing.

Radar energy will easily pass through ice and fresh water into the underlying sediment, revealing features on and below lake or river bottoms (Annan and Davis 1977; Fuchs et al. 2004; Jol and Albrecht 2004; Leckebusch 2003). Radar antennas can also be easily floated across the surface of a lake or river and directly on to the shore, all the while collecting data from the subsurface or even towed from a cable car over a river (Haeni et al. 2000). These techniques, however, will not work in salty or even slightly brackish water because the high electrical conductivity of the saline water will quickly dissipate the propagating electromagnetic energy as it enters the water column, leaving no energy to be transmitted to depth or reflected back to the surface. A few recent GPR surveys were conducted by hanging antennas from a low-flying helicopter. In this method, a distinct reflection was recorded from the ground and less distinct but still recognizable reflections from within the ground.

For expediency, during both continuous and step data acquisition, antenna transects acquired in a rectangular grid are usually collected in a sinuous pattern. One antenna transect is collected moving in one direction in the first transect and then in the opposite direction on the next parallel transect, offset some distance away. This collection pattern, with a standard transect offset, is then continued until all transects in a grid are acquired. Perpendicular cross-transects can be surveyed in the same fashion within the same grid, if necessary. If the reflection data are being stored digitally, simple computer programs can later reverse all the recorded traces for half of the acquired transects so that all reflection profiles produced within a grid have the same orientation with respect to a surveyed datum or baseline.

DATA RECORDING

For most standard GPR data collection, the elapsed time between radar pulse generation, reflection from interfaces in the ground, and final recording of the reflected wave at the receiving antenna is measured. The amplitude and wavelength of the reflected radar waves received back at the surface are also amplified, processed, and digitally recorded for immediate viewing on a computer screen and stored on some kind of digital medium for later postacquisition processing and display. Radar reflections are always recorded in *two-way time* because that is the time it takes a radar wave to travel from the surface transmitting antenna into the ground, be reflected off a discontinuity, and then travel back to the surface to be detected at another surface antenna and recorded.

Some of the earlier model GPR systems were only able to record reflection data on paper by means of a graphic recorder (Batey 1987; Fischer et al. 1980; Loker 1983). These systems use electrosensitive paper that moves across an electrically charged moving stylus, and all reflections in profiles are printed out on long rolls of paper as the antennas are moved along the ground in a transect. The stylus is fed the amplified incoming reflection data from the receiving antenna, with higher amplitudes printed as very dark shades of gray, while areas of little subsurface reflection remain white. A depth scale (in two-way travel time measured in nanoseconds) is also usually printed on the paper profiles. When using a graphic recorder, the operator can vary the speed of the paper and the sensitivity of the moving stylus to produce a wide variety of profile styles and exaggerations (Batey 1987; Fischer et al. 1980). This type of data recording has been almost totally superseded by digital recording systems, but a few of these systems are still in operation.

In other antiquated GPR units, reflected waves can be recorded on magnetic tape as small voltage changes around an arbitrary mean for later printout or digitization (Loker 1983). This kind of recorded data can be reprocessed later in order to convert the analog signal to digital data using a computer digitizer (Conyers 1995; Conyers and Spetzler 2002). Paper printouts of reflection profiles that are the only recorded archive of a GPR survey have even been digitized using a scanner, which assigns digital values to different shades of gray, approximating amplitudes of reflected waves in the ground. One would have to be very desperate to process an extremely important data set in this fashion, because it is not only laborious but also quite inexact.

In the early 1980s, GPR units were developed that recorded GPR reflections internally as digital data (Annan and Davis 1992; Geophysical Survey Systems 1987). With these units a computer is usually built into the control unit that allows reflection data to be easily processed, filtered, and spatially corrected as they are recorded. Digital units have now become the standard equipment in almost all GPR surveys, although good data can sometimes be acquired with the older analog units that have been converted to digital systems. Data are then saved to a computer hard drive or digital tape and most recently to flash memory chips.

ANTENNA VARIABLES

One of the most important variables in GPR surveys is the selection of antennas with the correct operating frequency for the depth necessary and the resolution of the features of interest (Huggenberger et al. 1994; Smith and Jol 1995). Commercial GPR antennas used in most archaeological applications range from 10- to 1,500-megahertz center frequency (Annan and Cosway 1994; Fenner 1992; Malagodi et al. 1996; Olson and Doolittle 1985; Jol and Bristow 2003). General-purpose GPR systems use dipole antennas that typically have a two-octave *band width*, meaning that the frequencies vary between one half and two times the center frequency. For example, a 300-megahertz center-frequency antenna generates radar energy with wavelengths ranging from about 150 to 600 megahertz. The frequency distribution of an idealized 500-megahertz wave, which is a bell-shaped distribution around a mean frequency, is shown in figure 3.9. In reality, depending on the electrical components and design of each individual antenna, the frequency distribution is rarely bell shaped but often an asymmetrical "spiky" distribution around a mean or median frequency. Figure 3.9 illustrates the actual frequency distribution derived from a radar pulse created from a 500-megahertz "bow tie" antenna that shows spikes in a number of frequencies with a mean frequency of about 505 megahertz. These variations in frequencies may be caused by irregularities in the antenna's surface (a bow tie–shaped copper plate), other electronic components located within the antenna system, or noise within the GPR system itself. Variations of this sort are common in all antennas, and each has its own irregularities, producing a different pulse signature and different dominant frequencies. Most important, just because a manufacturer identifies an antenna as having one frequency doesn't necessarily mean that it will produce radar energy with a center at exactly that frequency. If there is any question as to an antenna's frequency, a frequency distribution test should be performed that will yield a distribution curve like that shown in figure 3.9, prior to collecting data.

A primary goal of all antenna manufacturers is to produce a clean pulse of one wavelength in duration that can be transmitted into the ground. No antennas, however, produce perfectly clean pulses, and somewhat noisy reflection

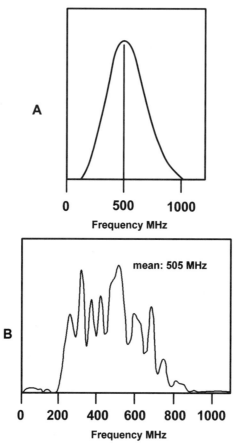

FIGURE 3.9
Antenna Frequency Distribution. Most antennas used in archaeology transmit at frequencies that vary about one octave around a mean center frequency. The theoretical distribution of energy transmitted from a 500-MHz antenna should take on a bell shape, with energy varying between 250 and 1,000 MHz (A). The actual frequency distribution of a 500-MHz antenna (B) when tested in the laboratory was found to have a mean of 505 MHz and a very uneven frequency distribution that varied between about 200 and 900 MHz.

records generated from noisy transmitted pulses are always the norm. A test was done on a 500-megahertz antenna (figure 3.10), and it was found that the transmitted pulse recorded in a noise-free environment is inherently noisy with the beginning of one strong pulse, recorded at 2 nanoseconds, followed by antenna "ringing" and other system noise after about 6 nanoseconds, making the transmitted energy one large pulse followed by many smaller-amplitude

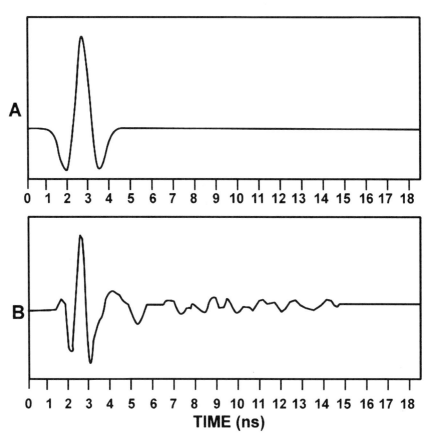

FIGURE 3.10
Radar Pulses. Antenna engineers strive to manufacture antennas that produce one pulse (A), but most antennas actually produce energy that is much noisier than desired due to imperfections in manufactured components and the addition of system and external noise (B).

waves. This antenna test is anything but the clean idealized pulse desired, which is almost always the case with commercially manufactured antennas. Postacquisition data processing can often "clean up" the noisy reflections recorded by most antennas.

Proper antenna center frequency selection can in most cases make the difference between success and failure of a GPR survey and must be planned for in advance. In general, the greater the depth of investigation, the lower the antenna frequency (longer transmitted wavelength) that is necessary (Smith and

Jol 1995). Lower-frequency antennas are for the most part much larger, heavier, and more difficult to transport to and within the field than high-frequency antennas. An older model 80-megahertz antenna used for continuous GPR acquisition is larger than a 42-gallon oil drum, cut in half lengthwise, and weighs between 125 and 150 pounds. Other newer-model low-frequency antennas can sometimes be moved by more than one person with the aid of tubular plastic supports (figure 3.11). These low-frequency antennas not only are difficult to transport to and from the field but must be moved along transect lines using some form of wheeled vehicle or sled. Often low-frequency antennas are so cumbersome that they must be used to collect data only in the step method, record-

FIGURE 3.11
Low-Frequency Antennas. Very low-frequency antennas are almost always unshielded and must be carried across the ground by one or more people. This antenna can produce center frequency energy of between 16 and 80 MHz depending on how it is configured.

ing one discrete reflection trace at each location as they are moved along a transect. In contrast, a 900-megahertz antenna is smaller than a shoe box, weighs very little, and can easily fit into a suitcase (figure 3.12).

Electrical engineers are continually modifying their antenna components and design parameters to more effectively produce "cleaner" and higher-power radar energy into the ground (Arcone 1995). Early GPR antennas, shaped like bow ties, were dipole antennas constructed of copper. In these antennas, an electrical current is applied to the center of a bow tie–shaped copper plate, which then radiates energy outward to its edges producing an electromagnetic field. One distinct pulse, like that shown in figure 3.10, is then created at the apex of the two arms of the bow tie, which propagates into space. To keep the electromagnetic field from oscillating many times within the antenna and creating many complex propagating waves instead of one distinct pulse, resistors are placed at the outer edges of the copper plate.

FIGURE 3.12
High-Frequency Antennas. A 900-MHz center frequency antenna system is very small and can be easily maneuvered across the ground.

Smaller, higher-frequency antennas are usually shielded, which allows energy propagation downward into the ground, but not upward or to the sides where it could be reflected off surface features, the antenna cables, or even the people pulling the antennas (Lanz et al. 1994). Shielding material that absorbs radar energy is usually placed on top and to the sides of the antenna, allowing energy to propagate only in a downward direction. When unshielded antennas are used, many reflections can be collected from objects on or above the ground surface and discrimination of individual targets in the ground can be difficult. However, if the unwanted reflections recorded from unshielded antennas all occur at approximately the same time—for instance, off a person pulling the antennas—then they can be easily filtered out later, if the data are recorded digitally (Leckebusch 2003). If reflections are recorded from randomly located trees, surface obstructions, or people moving about near the antenna, they usually cannot easily be discriminated from important subsurface reflections, and interpretation of the data is much more difficult unless noted specifically in field notes.

Most larger, lower-frequency antennas are more difficult to shield and therefore have the potential to receive many extraneous reflections from surface objects, and they can thus be quite "noisy." Bow tie antennas especially tend to "ring," which means that the oscillating electromagnetic field is not effectively dampened at the edges of the copper plate, producing a radar pulse that is less than perfect in its geometry. If a larger voltage were applied in an attempt to get more powerful electromagnetic energy into the ground, and produce a radar wave with higher amplitude, the antennas would tend to ring even more, producing a very noisy transmitted wave.

More recent antenna designs have abandoned the classic bow tie design and produce a cleaner radar pulse of higher amplitude, without extensive ringing. For instance, the GSSI 400-megahertz antenna creates a powerful, higher-amplitude pulse, which can travel deeper in the ground than the older 500-megahertz bow tie antenna, with greater subsurface resolution. The energy is also transmitted in a narrower beam as it moves in the ground, focusing more energy downward and creating "crisper" reflection profiles. Unfortunately, recent government regulations pertaining to radio bandwidth transmissions have necessitated that manufacturers produce antennas of lower power, negating some of these advances in antenna design. Fortunately, many of the newer antennas manufactured by most GPR firms are also smaller in size than earlier comparable models and therefore are easier to transport to and within the field.

RADAR PROPAGATION AND REFLECTION IN THE GROUND

Measurements of Radar Propagation and Reflection

The primary goal of most GPR investigations in archaeology is to differentiate and spatially map important subsurface interfaces. Any time radar energy crosses a contact between two materials in the ground with different physical or chemical properties, the velocity of the passing waves will change, and some energy will be reflected back toward the surface (figure 3.3). All sedimentary layers and other buried materials in the ground have particular properties that affect the velocity of electromagnetic energy propagation and therefore the strength of the reflected waves (Van Dam et al. 2002). Often the amount of reflection that occurs at buried interfaces is a function of differing retained or distributed water, which can be directly related to the physical properties of the buried units (Conyers 2004). The measurable properties of materials that affect radar propagation and reflection are electrical conductivity (related to the amount of retained water) and, to a lesser extent, magnetic permeability (Olhoeft 1981; Reynolds 1998: 688; Van Dam and Schlager 2000). If these are known (which is rarely the case for most sites, as detailed laboratory analyses must be conducted on soil and sediment samples), the amount of reflection at buried interfaces can be predicted.

Relative dielectric permittivity (RDP), also called the *dielectric constant*, takes the electrical and magnetic properties of buried materials into account and is a measure of the ability of a material to store a charge from an applied electromagnetic field and then transmit that energy (ASTM International 2003; von Hippel 1954; Wensink 1993). It is usually determined empirically from measurements in the field but can be directly measured in the laboratory, as will be discussed in chapter 5. In general, the greater the RDP of a material, the slower radar energy will move through it (figure 3.13). Relative dielectric permittivity is a general measurement of how well radar energy will be transmitted to depth. It therefore measures velocity of propagating radar energy and also its strength. For most archaeological applications, RDP values and measurements of the velocity of radar travel in the ground are used synonymously, as it is very difficult to measure or predict most of the other components of radar wave behavior used in the complex calculation of RDP.

For most archaeological studies, RDP and velocity are used interchangeably as a way to determine velocity of radar wave propagation in the ground. For instance, the RDP of fresh water is very high (about 80), but radar energy can easily be transmitted through it without being attenuated (especially when in a solid

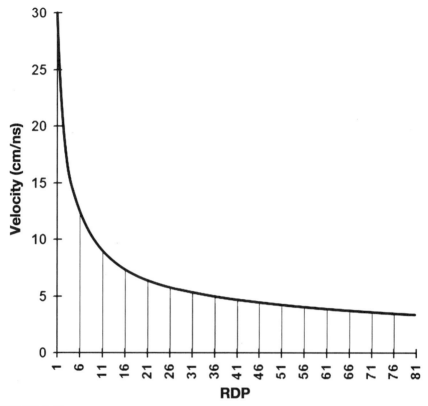

FIGURE 3.13
Graph of the Relative Dielectric Permittivity–Velocity Relationship. RDP is inversely related to radar travel velocity.

state—frozen), only at a very slow velocity. A bed of peat, which is composed almost wholly of organic material and fresh water, also has a high RDP but will also allow radar transmission to great depths, just at much slower speeds than in saturated sand or other materials (Clarke et al. 1999; Worsfold et al. 1986).

It is usually difficult to calculate RDP in the field, but it can be estimated using a number of field procedures discussed in chapter 5. It is always important to have some understanding of the RDP (or velocity) of the material in the ground at each site being studied, as it will be used to convert radar travel times to depth. Without this crucial understanding of how fast energy is traveling in the ground, many other preacquisition calibration procedures necessary for optimal data acquisition discussed in chapter 4 can be potentially compromised.

The relative dielectric permittivity of air, which exhibits only negligible electromagnetic attenuation, is approximately 1.0003 (Dobrin 1976) and is usually rounded to 1. Most soils and sediments found at archaeological sites have RDP values that range between 3 and about 25 (table 3.1). In a totally dry state, most naturally occurring materials in the ground have an RDP that varies little, usually between about 3 and 5. But if just a small amount of water is added to the material (which is almost always the case in natural conditions, even in the driest of deserts), the RDP will increase, sometimes dramatically (Conyers 2004). Take, for instance, sand (with an RDP of 3; see table 3.1) that is totally saturated with fresh water (with an RDP of 80). The overall RDP of this material can be estimated by taking the water-filled sand porosity of 30 percent and multiplying it by the RDP of water, which is 80. This is then added to the RDP of the sand (about 3) that makes up the other 70 percent of the total volume (calculated by taking 70 percent of 3). The RDP of the total volume would then estimated at 26.1: ([0.30 × 80] + [0.70 × 3]). In real field conditions, this easy calculation is

Table 3.1. Typical Relative Dielectric Permittivities (RDPs) of Common Geological Materials

Material	RDP
Air	1
Dry sand	3–5
Dry silt	3–30
Ice	3–4
Asphalt	3–5
Volcanic ash/pumice	4–7
Limestone	4–8
Granite	4–6
Permafrost	4–5
Coal	4–5
Shale	5–15
Clay	5–40
Concrete	6
Saturated silt	10–40
Dry sandy coastal land	10
Average organic-rich surface soil	12
Marsh or forested land	12
Organic-rich agricultural land	15
Saturated sand	20–30
Fresh water	80
Sea water	81–88

Modified from Davis and Annan (1989) and Geophysical Survey Systems (1987).

rarely possible, because almost all materials in the ground contain many other constituents such as clay and organic matter that can interact chemically or physically with the pore water, producing a complex blend of materials, each of which have different RDP values. Many volcanic and other hard rocks and moist sand and gravel can have RDPs that range from 6 to 16. Wet soils and clay-rich sediments or soils (Wensink 1993) often have RDPs which can approach 40 or 50 (table 3.1). In unsaturated sediment, with little or no mineralogical clay, RDPs are usually 5 or lower.

The RDP values shown in table 3.1 are very approximate and can vary greatly over a site and with depth of burial. If data about the types of material in the ground are not immediately available, the RDP of the ground can only be estimated using a number of field methods discussed in chapter 5. Equation 3.1, which relates RDP and radar velocity of a material, is shown below:

$$K = \left(\frac{C}{V}\right)^2$$

K = relative dielectric permittivity (RDP) of the material through which the radar energy passes
C = speed of light (.2998 meters per nanosecond)
V = velocity of the material through which the radar passes (in meters per nanosecond)

EQUATION 3.1
Relative dielectric permittivity and radar velocity relationship in GPR.

To generate a significant radar reflection, the change in RDP between two bounding materials must occur over a short distance. When the RDP changes gradually with depth, only small differences in reflectivity will occur every few centimeters in the ground, and very weak or no reflections at all will be generated (Van Dam and Schlager 2000). The amplitude of reflections generated at an interface between two materials with known RDPs can be calculated using equation 3.2 (Sellmann et al. 1983; Walden and Hosken 1985; Van Dam et al. 2002). But the inability to precisely measure the physical and chemical parameters of buried units in the field usually precludes accurate calculations of specific amounts of reflectivity in most archaeological contexts, and usually only estimates can be made.

$$R = \left(\sqrt{K_1} - \sqrt{K_2}\right)\left(\sqrt{K_1} + \sqrt{K_2}\right)$$

R = coefficient of reflectivity at a buried surface
K_1 = RDP of the overlying material
K_2 = RDP of the underlying material

EQUATION 3.2
The coefficient of reflectivity at an interface between materials of differing relative dielectric permittivity. The higher the coefficient, the higher the amplitude of reflected waves at the interface.

The highest-amplitude radar reflections usually occur at an interface of two relatively thick layers that have greatly varying properties. For instance, a difference of this sort might be between a compacted clay floor of a buried pit house and overlying sand or gravel layer that covers it (e.g., figure 2.1). Often other important stratigraphic interfaces or buried archaeological features of interest will also produce high-amplitude reflections, but not always. If the target archaeological features are composed of almost exactly the same substance as the material surrounding them, or if those materials have about the same physical and chemical properties, there will no variation RDP between them, and little or no reflection will occur at their interface. Calculations other than equation 3.2 that can be used to quantify reflectivity, and therefore the amplitudes or reflected waves at interfaces, are discussed in greater detail later.

Dispersion and Attenuation of Radar Energy in the Ground

Another factor that affects the depth of penetration and amplitude of reflected radar waves in the ground is dispersion and energy attenuation. These occur because most ground is at least slightly electrically conductive, which dissipates and absorbs propagating waves as they move through it. As energy moves more deeply in the ground, less is therefore available for reflection and amplitudes of reflections at buried interfaces also decrease (figure 3.3). When what remains of the original transmitted energy is finally reflected back toward the surface from deep within the ground, that remaining energy will suffer additional attenuation within the materials through which it passes before finally reaching the receiving antenna. As a result, radar energy is always progressively weakened as it moves through the ground. Reflections must therefore be generated at subsurface interfaces that have sufficient RDP contrast and also must be located at a shallow enough depth where sufficient radar energy is still available to be reflected. Sometimes filtering and wave amplification techniques, which will be discussed in chapter 6, can be applied to very weak reflections after acquisition that can enhance their very low amplitudes in order to make them more visible. However, there is always a maximum depth of radar energy penetration and therefore potential reflection, which is different for every site, no matter what antenna frequency is used or postacquisition processing techniques are employed.

The maximum effective depth of penetration of GPR waves is a function of the frequency of the waves that are propagated into the ground and the physical characteristics of the material through which they pass (Annan et al. 1975; Batey 1987; Geophysical Survey Systems 1987; Keller 1988; Olhoeft 1981). Soils, sediment, or rocks that are termed *dielectric* will permit the passage of a great deal of electromagnetic energy without dissipating it. The more electrically conductive a material, the less dielectric it is, and the larger amount of energy will attenuate at a shallower depth. In a highly conductive medium, the electrical component of the propagating electromagnetic wave is rapidly conducted away, and when this happens, the wave as a whole dissipates. This occurs because for propagation to occur, the electrical and magnetic waves must constantly "feed" on each other during transmission (figure 3.1).

Highly electrically conductive media include those that contain salt water and some that have certain types of electrically conductive clay, especially if that clay is wet. Any soil or sediment that contains soluble salts or electrolytes in the groundwater will also create a medium with a high electrical conductivity. Agricultural runoff that is partially saturated with soluble nitrogen and potassium can potentially increase the electrical conductivity of a medium, as will moist calcium carbonate impregnated soils in arid regions. Often desert soils, even if they appear to be extremely dry and therefore should readily allow radar transmission, contain hydrous salts in their interstices, which conduct electricity. In these types of soils, radar energy will often become attenuated at a shallow depth.

Soil chemistry, especially the types and structures of different clay minerals, also plays a role in radar energy transmission, but this mechanism is still poorly understood (Rhoades et al. 1976; Walker et al. 1973; Wensink 1993). Some common soils are composed of mineralogic clays with a high ionic displacement such as montmoriollionite, smectite, and bentonite (Birkeland 1999). These are three-layer clays that are generally termed swelling clays. All have the ability to both hold water in their atomic structure (and therefore swell when wet). Their molecular structure, when it contains water, allows the easy movement of ions, making them good electrical conductors. In comparison, two-layer clays such as kaolonite and three-layer nonswelling clays such as illite are relatively resistive, and their presence in a soil, even when wet, will often not greatly impede radar energy transmission.

It used to be assumed that wet clay, no matter what type, would attenuate radar waves, and it was therefore unsuitable for GPR surveys (Leckebusch 2003). While this is often the case, good radar reflections have been recorded at a depth of more than 2 meters with a 400-megahertz antenna in western Oregon, in ground composed almost entirely of saturated clay (figure 3.14). This unusual success was a mystery at the time, and only after returning from the field and analyzing the soil and sediment samples collected at the site was it discovered that the clay was not mineral clay. The ground was actually sediment composed of rock fragments that were of clay size, which had not yet undergone diagenesis into mineral clays (Birkeland 1999); therefore, the material did not have the high electrical conductivity properties of many clay minerals discussed earlier. How one would readily determine in the field whether clay at a survey area was mineral clay or just sediment composed of clay-sized rock fragments, without detailed sedimentological and mineralogical analysis, is unknown.

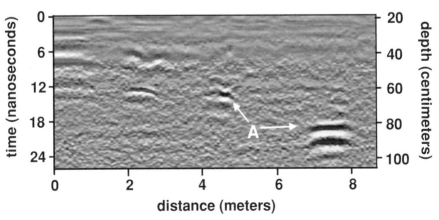

FIGURE 3.14
Reflection Profile in Wet Clay. Good radar reflections (A) were recorded at more than 25 nanoseconds in a homogeneous saturated clay in western Oregon. Laboratory tests of this material showed it to have a low conductivity but a high RDP, and therefore energy traveled within it with little attenuation, but at a relatively slow velocity.

It was also interesting that this saturated clay, with a porosity of about 40 percent, had a very high RDP, because of the amount of retained water. Radar energy therefore readily traveled within it (but at a low velocity) because it was a relatively nonconductive medium. In contrast, the same 400-megahertz antenna was used to collect data at a site in southern Colorado, in extremely dry sandy silt. The maximum depth of radar penetration there was only about 50 centimeters (figure 3.8). Only after the GPR survey was complete, and a mineralogical analysis of the sediment made, was the presence of bentonite clay found in the medium. This water-holding clay made the otherwise dry sediment electrically conductive, explaining the rapid radar energy attenuation at such a shallow depth.

Another similar survey in coastal Peru, with presumably dry sandy soil (it had not rained in that area for decades), also had a very high electrical conductivity, and therefore radar energy was also attenuated at a very shallow depth. In this case, electrically conductive salts were bound with clay in the sand, making the medium as a whole highly electrically conductive and therefore nondielectric.

These examples illustrate how the old GPR adage that dry sandy soils are good for radar penetration while wet clay is bad can be very misleading, as there are many other more important factors controlling radar propagation (Conyers 2004; Gerber et al. 2004). The most important is electrical conductivity, which is often difficult to predict in advance of conducting a GPR survey.

Undecomposed organic matter, such as peat, is often relatively nonconductive, even when wet. The high percentage of water in these sediments, however, will drastically slow radar travel times. Peat will therefore have a high RDP (low propagation velocity), while still allowing radar wave transmission to sometimes great depths (Clarke et al. 1999; Leopold and Volkel 2003). Decomposing organic matter, however, sometimes accumulates metals, especially in a chemically reducing environment, and those metals can increase the material's overall electrical conductivity (Van Dam et al. 2002). This type of wet organic material will also increase the overall acidity in the ground, creating mobile hydrogen ions, which also allow the greater passage of electricity and an increase in the conductivity in some cases.

Other minerals in the ground, especially those that can dissolve in water, will create free ions, which allow for greater electrical conductivity. Sulfates, carbonate minerals, iron, salts of all sorts, and any charged elemental species of mineral will create a highly conductive ground and readily attenuate radar energy at shallow depths (Van Dam et al. 2002). Under the very unfavorable conditions of wet (with slightly saline water), calcareous sediment or soils that contain certain clay-rich minerals, the maximum depth of GPR penetration in the ground can be much less than a meter.

Barring detailed soil chemistry studies at sites prior to data acquisition, the best method of determining an area's conduciveness for GPR studies is to collect GPR data and visually determine depth of energy penetration in reflection profiles. Some researchers have attempted to use electromagnetic conductivity meters (EMs) to measure near-surface ground conductivity, as those readings will generally determine whether GPR energy will successfully be transmitted to the depth desired. Care must be taken using this method, because EM tools must be correctly calibrated to a known media first, or conductivity measurements can be invalid. This method also necessitates a trip to the site first for EM data collection. If GPR equipment were also available, it would probably be better to just collect a radar profile or two at a prospective site to see how deeply energy will penetrate, which would yield more definitive results.

Research is still ongoing to devise an instrument that will quickly and accurately measure soil conductivity of samples, which can then be used to determine the efficacy of GPR in a survey area prior to going to the field. Some have attempted to use devices that were developed to determine the moisture content in grain shipped to storage elevators, which also use dielectric methods. Others have resorted to simple direct current devices that pass an electrical current from one electrode to another in the ground, measuring electrical resistivity, which is the inverse of conductivity. Both have had marginal success in predicting radar transmission because they are measuring either just one sample, which may not be indicative of the ground as a whole, or an electrical current whose pathway in the ground cannot be readily determined. To date, there is no really good way to accurately measure electrical conductivity in the ground as it affects radar transmission, outside of some calibrated EM devices and laboratory tests of samples.

Magnetic permeability also affects radar penetration depth in a medium. It is a measure of the ability of a medium to become magnetized when an electromagnetic field is imposed upon it (Sheriff 1984). Most soils and sediments are only slightly magnetic and therefore usually have a low magnetic permeability. The higher the magnetic permeability, the more electromagnetic energy will be attenuated during its transmission. When this occurs, the magnetic portion of the EM wave is destroyed, just as when electrical conductivity is increased, the electrical component is lost. Media that contain magnetite minerals, iron oxide cement, or iron-rich soils can all have a high magnetic permeability and therefore often transmit radar energy poorly (Van Dam et al. 2002). Poor radar energy penetration caused by this physical property has been encountered in basaltic sands in Hawaii (Olhoeft 1998) and unweathered granite outwash sediments in Arizona, both of which contain large amount of magnetite and other slightly magnetic iron-rich minerals. But in Iceland basalt seems to be capable of transmitting energy to many meters with low-frequency antennas (Cassidy et al. 2004).

Radar energy will not penetrate metal. A metal object will reflect all of the radar energy that strikes it and will "shadow" anything directly underneath. Buried metal objects are quite easy to see in GPR reflection profiles because they usually create multiple reflections stacked on top of one another below the metal object (figure 3.15). This is caused by radar energy reflecting off the metal object, traveling back to the ground surface to be reflected again from the ground–air interface, back to the metal object, and then again to the surface. When this occurs, many reflections of this sort are often stacked on top of each other, a good indicator of buried metal. Other materials besides metal

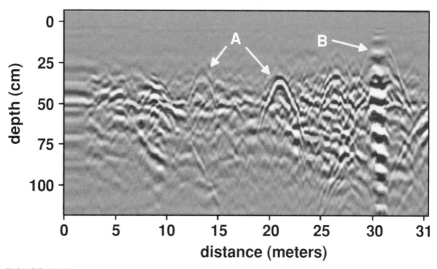

FIGURE 3.15
Point Source Hyperbolas. Buried pipes (A) have generated reflection hyperbolas in profile. The hyperbola on the right was generated from a metal pipe and the lower-amplitude hyperbola on the left from a plastic pipe. The series of high-amplitude reflections that are stacked vertically at location B were generated by a large piece of metal near the ground surface.

are also highly reflective, such as baked clay or some plastic objects, so multiple stacked reflections as shown in figure 3.15 do not indicate solely metal objects.

The depth of radar energy penetration and subsurface resolution is actually highly variable, depending on many site-specific factors such as overburden composition, porosity, and the amount of retained moisture. It is important to remember that in ground conditions that are highly conductive, radar energy will become attenuated at shallow depths, no matter what its wavelength. There is a common misconception that if a high-frequency antenna (say, a 500-megahertz) is only capable of transmitting energy to about 50 centimeters in the ground, then a lower-frequency antenna will transmit deeper. If a 300-megahertz antenna was also tried, and its maximum depth of penetration was about the same, then the ground is almost surely highly electrically conductive, and no antenna, no matter what frequency or how powerful, would be able to transmit to a greater depth.

At very high frequencies, usually greater than 1,500 megahertz, some geologic materials containing water will exhibit higher than normal energy attenuation due to energy loss from molecular relaxation (Annan and Cosway 1994; Olhoeft 1994b). This is usually not a problem in most archaeological applications, but due to the wide bandwidth of commercial antennas, some of the radar energy produced from center-frequency antennas of 800 megahertz or greater might be affected. Molecular relaxation occurs because water molecules are bipolar and will rotate and become aligned within an imposed electromagnetic field. This rotation causes the radar energy to be converted to mechanical energy, which is then dissipated as heat, much like how a microwave oven works (ASTM International 2003). Energy loss of this sort, which is frequency-dependent, is referred to as *dielectric relaxation,* and usually becomes a factor only if a GPR survey is being conducted in a very wet environment with a high-frequency antenna, which is rare.

Reflection Types

A series of reflection traces collected along a transect that are produced from a buried layer will generate a horizontal or subhorizontal line in profiles (either dark or light in gray scale reflection profiles) that is referred to simply as a *planar reflection* (as in the pit house floor in figure 2.1). These types of distinct reflections are usually generated from a subsurface boundary such as a stratigraphic horizon or some other physical discontinuity such as the water table, a buried soil horizon, or a horizontal feature of archaeological interest. There can also be *point source reflections* that are generated from one distinct aerially restricted feature or object in the subsurface (figure 3.15). The buried materials that generate these types of point source reflections could be individual rocks, metal objects, pipes that are crossed at right angles, and a great variety of other smaller things of this sort. They are visible in two-dimensional profiles as reflection hyperbolas, even though they were generated from a "point," or aerially restricted feature in the ground. A large number or density of hyperbolas in a reflection profile can often make interpretation difficult because many closely spaced hyperbolic reflections produce very complex and "busy" profiles (figure 3.16).

Point source reflection hyperbolas, sometimes termed *diffractions*, are generated because most GPR antennas produce a transmitted radar beam that propagates downward from the surface in a conical pattern, radiating outward as energy travels to depth (figure 3.17). Radar energy will therefore be reflected

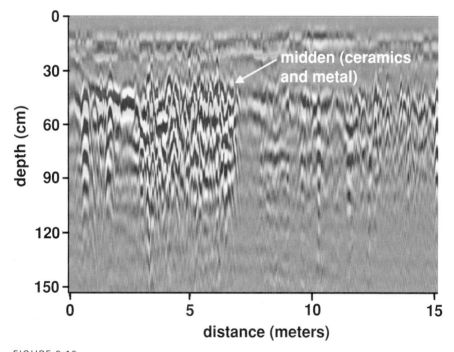

FIGURE 3.16
Many Small Hyperbolas. A cluttered group of many hyperbolas can be generated from small artifacts, such as these that were generated by small pieces of metal and broken ceramics, creating a complex reflection profile. This feature was excavated along the San Gabriel River in southern California, and found to be a nineteenth-century household midden.

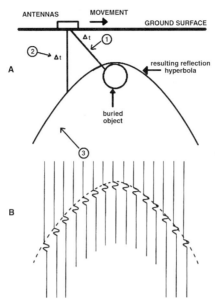

FIGURE 3.17
Generation of a Reflection Hyperbola. The conical projection of radar energy into the ground will allow radar energy to travel in an oblique direction to a buried point source (1) as seen in A. The two-way time (Δt) is recorded and plotted in depth directly below the antenna where it was recorded (2). When many such reflections are recorded as the surface antennas move toward and then away from a buried object, the result is a reflection hyperbola (3), when all traces are viewed in profile, as seen in B.

from buried features that are not located directly below the transmitting antenna but are still within the "beam" of propagating waves. Oblique radar wave travel paths to and from the ground surface are longer (as measured in radar travel time), but reflections generated from objects not located directly below the antennas will still be recorded as if they were directly below, but just deeper in the ground. As the surface antenna moves closer to a buried point source, the receiving antenna will continue to record reflections from the buried point source prior to arriving directly on top of it and continue to "see" it after it has passed. A reflection hyperbola is then generated because the time it takes for the energy to move from the antenna to the object along many oblique paths is greater the farther the antenna is away from the source of the reflection. As the antenna moves closer to the buried object, the reflection from it is recorded closer in time until the antenna is directly on top of it. The same phenomena is repeated in reverse as the antenna passes away from the source, resulting in a hyperbola where only its apex denotes the actual location of the buried reflection source, with the arms of the hyperbola creating a record of reflections that traveled the oblique wave paths. In some cases, only half of a hyperbola may be recorded, if just the corner or edge of a more planar feature is causing a discrete reflection, such as the edge of a buried house floor or platform of some sort.

The presence of reflection hyperbolas is considered by some geophysicists to be a distraction during data interpretation because they are not denoting the "real" location of buried features but are the product of the complex geometry of radar wave travel paths in the ground. Their presence, however, can aid in interpretation because hyperbolas are easily identified in reflection profiles and denote a specific size and possible geometry of objects in the ground. Most important, their utility in determining velocity (discussed in chapter 5) cannot be overemphasized. If it later becomes necessary to remove reflection hyperbolas from profiles, software programs (discussed in chapter 6) can "collapse" the hyperbola arms to their apex using data migration procedures. This is often necessary when performing more complex spatial analysis and mapping buried features in three dimensions where many hyperbola "tails" can create false anomalies, tending to blur the location of some objects in maps and three-dimensional images.

Resolution of Subsurface Features

Subsurface resolution is mostly a function of the wavelength of propagating radar energy and the geometry of the buried materials in the ground of interest. Low-frequency antennas (those of 10 to 120 megahertz) generate long-wavelength radar energy that can penetrate up to 50 meters or more in certain conditions but are capable of resolving only very large subsurface features. In pure ice, antennas of this frequency have been known to transmit radar energy many kilometers, and they are commonly used to determine the thickness of glacial ice or the orientation of subice bedrock surfaces (Bogorodsky et al. 1985; Delaney et al. 2004). In contrast, the penetration depth of a 900-megahertz antenna is about 1 meter, and often less, in typical ground conditions, but its generated reflections can resolve features down to a few centimeters in diameter. A trade-off therefore exists between depth of penetration and subsurface resolution. Table 3.2 shows the dominant wavelength for different center-frequency antennas, and how those wavelengths change as energy moves through materials of differing RDP.

The ability to resolve buried features is largely a function of the wavelength of energy reaching them at the depth they are buried. A "rule of thumb" is that the minimum object size that can be resolved is about 25 percent of the downloaded wavelength reaching them in the ground. *Downloading* of radar energy always occurs as energy passes in the ground (Jol and Bristow 2003) and decreases in frequency, increasing the propagating wavelength of the radar waves. For instance, a 400-megahertz center-frequency antenna will generate energy with a wavelength of 75 centimeters in air (table 3.2). When that energy downloads as it moves into the ground, its dominant frequency decreases to about 300 megahertz, which is a wavelength of about 100 centimeters. It is almost impossible to calculate what the downloaded wavelengths of any radar energy transmitted from an antenna would be, and it is usually sufficient to be aware of the downloading phenomenon.

Determining what the wavelength of any frequency radar wave might be in the ground is further complicated by additional changes in wavelength as energy passes through materials with different RDPs. In all cases, radar waves moving in the ground will decrease their wavelength from the downloaded energy. For instance, a 300-megahertz antenna generates a wavelength of about 1 meter in air (table 3.2). This frequency energy would download to about 200 megahertz or so in the ground, which would produce wavelengths of propagating waves of about 1.5 meters. But as the energy passes progressively through materials with increasing RDP (say, an RDP of 5), those same radar waves would then decrease their wavelength to about 0.67 meters (table 3.2). If they were then to travel even deeper in the ground and pass through material of even higher RDP, the wavelength would likely decrease even further. Although difficult to calculate and know for sure, an estimation of these wavelength changes is important because an object much smaller than about 25 percent of the wavelength of radar energy intersecting them would in all likelihood not be resolvable in reflection profiles. In the prior example, therefore, an object smaller than about 25 percent of 0.67 meters (about 17 centimeters in diameter) would probably not be resolvable using a 300-megahertz antenna.

Unfortunately, it is often not known in advance what the target depth of archaeological features of interest is, their dimensions, or often the ground conditions and their physical properties. Most important, the ability to transmit radar energy to the depth necessary is often not known until one actually collects some reflection profiles. The best one can usually do prior to going to the field is to make some rough calculations from the best knowledge available, and then take the antennas that will hopefully be necessary for the task. Antenna choice can therefore be a difficult decision. As a general rule, if the target features are within about 1 meter of the ground surface, antennas between 400 and 900 megahertz will be adequate to transmit energy to that depth and to resolve most features and associated stratigraphy (table 3.2). If target features are small, and greater resolution is needed, the higher frequency antennas in this range should be used. If the target depth is between 1 and 3 meters, antennas from 500 to 200 megahertz or so are probably optimal. Radar energy with a frequency higher than about 500 megahertz will rarely transmit energy to greater than 2 meters in the ground, except in exceptionally dielectric media. Targets buried deeper than about 3 or 4 meters will require antennas with frequencies lower than 200 megahertz, but it is important to remember that the wavelengths generated from these frequencies are only capable of resolving fairly large features. Also, the deeper in the ground the energy must penetrate, the more spreading of the transmission

Table 3.2. Wavelength (in meters) of Radar Waves in Media of a Given RDP and Frequency

	Frequency (MHz)									
RDP	100	200	300	400	500	600	700	800	900	1,000
1 (air)	2.998	1.499	0.999	0.750	0.600	0.500	0.428	0.375	0.333	0.300
2	2.120	1.060	0.707	0.530	0.424	0.353	0.303	0.265	0.236	0.212
3	1.731	0.865	0.577	0.433	0.346	0.288	0.247	0.216	0.192	0.173
4	1.499	0.750	0.500	0.375	0.300	0.250	0.214	0.187	0.167	0.150
5	1.341	0.670	0.447	0.335	0.268	0.223	0.192	0.168	0.149	0.134
6	1.224	0.612	0.408	0.306	0.245	0.204	0.175	0.153	0.136	0.122
7	1.133	0.567	0.378	0.283	0.227	0.189	0.162	0.142	0.126	0.113
8	1.060	0.530	0.353	0.265	0.212	0.177	0.151	0.132	0.118	0.106
9	0.999	0.500	0.333	0.250	0.200	0.167	0.143	0.125	0.111	0.100
10	0.948	0.474	0.316	0.237	0.190	0.158	0.135	0.119	0.105	0.095
11	0.904	0.452	0.301	0.226	0.181	0.151	0.129	0.113	0.100	0.090
12	0.865	0.433	0.288	0.216	0.173	0.144	0.124	0.108	0.096	0.087
13	0.831	0.416	0.277	0.208	0.166	0.139	0.119	0.104	0.092	0.083
14	0.801	0.401	0.267	0.200	0.160	0.134	0.114	0.100	0.089	0.080
15	0.774	0.387	0.258	0.194	0.155	0.129	0.111	0.097	0.086	0.077
16	0.750	0.375	0.250	0.187	0.150	0.125	0.107	0.094	0.083	0.075
17	0.727	0.364	0.242	0.182	0.145	0.121	0.104	0.091	0.081	0.073
18	0.707	0.353	0.236	0.177	0.141	0.118	0.101	0.088	0.079	0.071
19	0.688	0.344	0.229	0.172	0.138	0.115	0.098	0.086	0.076	0.069
20	0.670	0.335	0.223	0.168	0.134	0.112	0.096	0.084	0.074	0.067
30	0.547	0.274	0.182	0.137	0.109	0.091	0.078	0.068	0.061	0.055
40	0.474	0.237	0.158	0.119	0.095	0.079	0.068	0.059	0.053	0.047
50	0.424	0.212	0.141	0.106	0.085	0.071	0.061	0.053	0.047	0.042
60	0.387	0.194	0.129	0.097	0.077	0.065	0.055	0.048	0.043	0.039
70	0.358	0.179	0.119	0.090	0.072	0.060	0.051	0.045	0.040	0.036
80	0.335	0.168	0.112	0.084	0.067	0.056	0.048	0.042	0.037	0.034

beam, and the more energy attenuation, so it may not be possible to get good reflections from deeper than 3 or 4 meters in most ground conditions, no matter what the antenna frequency. If targets are buried deeper than 5 meters or so, antennas with frequencies lower than about 100 megahertz are usually necessary, and ground that is dielectric enough to allow radar penetration to that depth is uncommon. Usually it must be exceptionally dry and lacking in conductive clay or salts, dry unweathered volcanic ash or perhaps permafrost, or very deeply frozen soil.

Another way to determine if features of a certain size are resolvable in reflection profiles is to calculate how much of a propagating beam of radar energy will illuminate them. As a basic guideline, the cross-sectional area of the target to be detected should approximate the size of the energy illumination pattern at the target depth or a little smaller (figure 3.18). If the radiating antenna is properly shielded so that energy is being propagated in a mostly downward direction, this elliptical *illumination pattern* (also called the *footprint*) on a horizontal surface (figure 3.18) can be calculated (Annan and Cosway 1992). If the target is much smaller than the footprint size, then only a fraction of the transmitted energy that intersects it within the cone of transmission will be reflected to the surface. The small number of reflections returned from a very small buried feature in this case would probably be indistinguishable from background reflections generated elsewhere within the transmission cone and therefore be invisible in reflection profiles. Small features of this sort might still be detectable if they create very high-amplitude reflections or after raw reflection data are computer processed to increase resolution, discussed in chapter 6.

An estimation of footprint size is also important when designing transect spacing within a grid so that all subsurface features of importance are illuminated by the transmitted radar energy and can therefore generate reflections. In general, the angle of the transmission cone, and therefore the size of the footprint, varies as a function of the relative dielectric permittivity of the material through which the waves pass, and the frequency of the radar energy emitted from the antenna. Equation 3 in figure 3.18 can be used to estimate the width of the transmission beam at varying depths. This equation can only be used as a rough approximation of real-world conditions because it assumes a consistent dielectric permittivity of the medium through which the radar energy passes and one single antenna frequency. Outside strictly controlled laboratory conditions, this is never the case. Sedimentary and soil layers within the ground almost always have variable chemical constituents, differences in retained moisture, compaction, and porosity. These and other variables create a complex layered system with varying relative dielectric permittivities and therefore energy transmission patterns, which are often difficult to define precisely.

Higher-frequency antennas, such as the 900-megahertz or higher, have quite narrow cones of propagation, while the 200- and 300-megahertz-frequency antennas can spread energy outward a meter or more at depths of only about one or two meters below the ground surface (figure 3.18). The cones of radar transmission in the ground are in reality more elliptical than circular in geometry because the electrical field produced by the antenna is generated parallel to its long axis and is therefore usually radiating into the ground perpendicular to the direction of antenna movement along the ground surface. If the antenna dipoles are positioned perpendicular to the direction of movement along

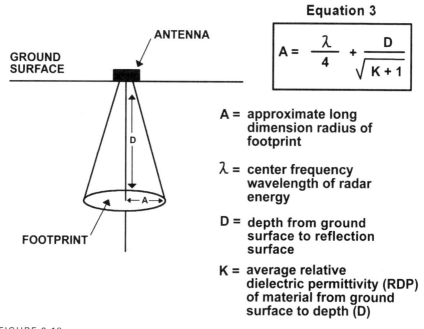

Equation 3

$$A = \frac{\lambda}{4} + \frac{D}{\sqrt{K+1}}$$

A = approximate long dimension radius of footprint

λ = center frequency wavelength of radar energy

D = depth from ground surface to reflection surface

K = average relative dielectric permittivity (RDP) of material from ground surface to depth (D)

FIGURE 3.18
Conical Spreading of Radar Energy in the Ground. Radar energy spreads out in a conical projection as it travels into the ground. The approximate size of the radiation footprint at a depth in the ground can be estimated from the antenna frequency and the RDP of the ground through which the energy passes.

a transect (the usual orientation), the cone of propagation is more elongated parallel to the direction of transport. This will cause a greater propagation outward both in front and behind the antennas, and less to the sides.

The illumination footprint is much larger in dimension when radar energy travels through a material with a low RDP (figure 3.19). Higher-RDP material tends to focus the beam of transmission, decreasing the radius of the subsurface footprint. Therefore, when conducting a survey in ground with a high RDP, transect lines should be more closely spaced in order to make sure all subsurface features are illuminated with (and therefore can potentially reflect) radar energy.

Any estimation of the orientation of transmitted energy is also complicated by the knowledge that radar energy propagated from most surface antennas is not one distinct frequency but can range many hundreds of megahertz around

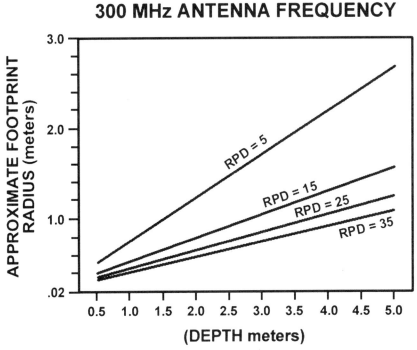

FIGURE 3.19
Radiation Footprint Differences with Differing Ground RDP. The cone of transmission is much broader (and therefore the footprint radius greater) with depth when the RDP of the material is low. In high-RDP material, the transmission cone is narrower, and its footprint radius at any one depth is much smaller.

a center frequency. If one were to make a series of calculations on each layer in the ground (assuming all the soil and sediment physical and chemical variables could be quantified), and if one distinct antenna frequency was assumed, then the "cone" of transmission would be seen to widen in some layers, narrow in others, and create a very complex three-dimensional pattern. The best one can usually do for most archaeological applications is to estimate the radar beam geometry and footprint size based on approximate field conditions and the center frequency of the antenna to be used. Some determination of the propagation beam dimensions, however, is always important prior to conducting a survey so that grid lines can be spaced at distances smaller than the maximum footprint dimension at the depth necessary to delineate the features of interest (equation 3 in figure 3.18). Any wider spacing of survey transects may allow important subsurface features to go undetected.

Recent field studies have shown that the amount of radar energy emitted from an antenna is greater directly under the antenna, and it tends to decrease in the more splayed portion of the cone of radiation (Leckebusch 2003; Neubauer et al. 2002). For this reason, survey line transects should be spaced as closely together as field conditions, equipment, and time allow (Jol and Bristow 2003). Comparisons of maps produced in grids containing variously spaced transects indicate that the highest-resolution images of buried features will always result from reflection data acquired in more closely spaced transects (Conyers et al. 2002; Neubauer et al. 2002).

Resolving a sequence of buried horizontal surfaces in the ground is even more complicated than determining whether individual objects might be visible in reflection profiles. To distinguish radar reflections generated from two parallel buried layers (e.g., the top and bottom of a large planar object), the two interfaces must be separated by at least one wavelength of the energy that is encountering them (Davis and Annan 1989). If the two reflections generated are not separated by that distance, then the resulting reflected waves from both the top and bottom will be either destroyed or unrecognizable due to constructive and destructive interference, as illustrated in figure 3.20. When two interfaces are separated by greater than one wavelength, two distinct reflections are generated from each interface, and both the top and bottom of the feature can potentially be resolved.

If only one buried planar surface is being mapped, then the first arrival reflected from that interface can usually be resolved, independent of the wavelength. Reflections generated from buried surfaces using longer-wavelength

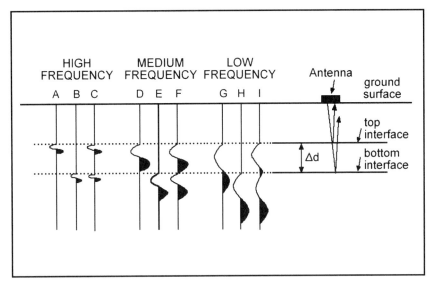

FIGURE 3.20
Resolution of Interfaces. Depending on the frequency of energy transmitted into the ground and the distance between two planar interfaces (Δd), reflections from the top and bottom of a layer may or may not be visible in a reflection profile. High-frequency energy will generate a small enough wavelength so that the top (A) and bottom (B) will produce a reflection, and the composite reflection trace of the two (C) can define both interfaces. Medium-frequency antennas with a longer wavelength will just barely have enough definition from the top and bottom (D and E) to produce a composite reflection trace (F) that exhibits both interfaces. Low-frequency antennas may produce a wave that will reflect off both interfaces (G and H), but the composite reflection trace is affected by constructive and destructive interference of the two waves, and only the top of the interface is visible in the composite reflection trace (I).

radar energy tend to be less sharp than those from higher-frequency antennas, when viewed in a standard reflection profile (Annan and Cosway 1992). This is because the longer wavelength energy tends to spread out more as it travels in the ground (figure 3.19) and is therefore reflected from more surface area (a larger footprint) than would occur with higher-frequency energy. As a result, small irregularities on the buried surface would not be visible, as they would likely be "averaged out" in the recorded reflection traces derived from a wider cone of transmission. This results in an averaging of many reflections from the buried surface and a more blurry composite reflection generated from the buried interface. In contrast, a high-frequency antenna produces a transmission cone that is much narrower, and its resolution of subsurface features on the same buried surface will be much greater as reflections received at the ground surface were generated from much less of the buried surface area. The trade-off, however, is that the higher-frequency waves will not propagate as far into the ground. This phenomenon is demonstrated in figure 3.21 where differing buried stratigraphic horizons are visible in both 300- and 500-megahertz center-frequency reflection data along the same traverse. A very different set of buried layers is detectable in both profiles because of variations in the depth of energy propagation, the resolution of those buried interfaces due to the wavelength differences of the energy reflected, and the footprint size on the reflection surfaces. In the step collection method the spacing of traces also affects subsur-

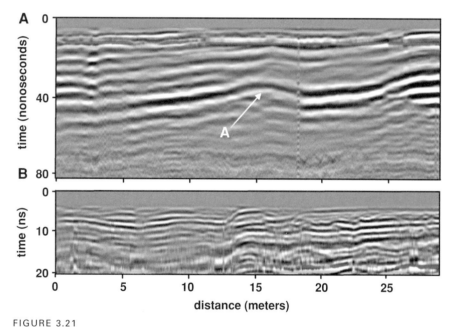

FIGURE 3.21
Resolution of Stratigraphy as a Function of Frequency. A 300-MHz reflection profile (profile A) generated energy that penetrated to about 50 nanoseconds and reflected off a number of subsurface interfaces. It exhibits relatively poor stratigraphic resolution in the upper 10 nanoseconds or so. In contrast, a 500-MHz profile along the same transect (profile B) exhibits high-definition reflections to only about 20 nanoseconds, but with good shallow resolution, showing many stratigraphic layers not visible in the lower-frequency profile. The upward bowing layer in profile B produced a lower-amplitude reflection along its crest (A) because some energy was scattered away from the surface antenna and was not received back at the surface antenna.

face resolution with resolution decreasing with greater spacing (Jol and Bristow 2003).

Radar energy that is reflected off a buried interface that slopes away from a transmitting antenna will probably not travel back to the paired receiving antenna on the ground surface. In this case, all reflected energy would be lost, and the sloping interface will go unnoticed in reflection profiles. A buried surface with this orientation would only be visible if an additional antenna traverse were located in an orientation where that same buried interface is sloping toward the receiving antenna. This is one reason why it is important to always acquire transects of data within a closely spaced surface grid, and sometimes in traverses perpendicular to each other, in order for all buried features, no matter what their orientation, to be visible. This is the same concept exploited in the "stealth" technology for airplane construction. Wings and the fuselage of stealth aircraft are built in a geometry that reflects energy in any direction but back to the receiving antenna, making them essentially invisible to radar.

In most geological and archaeological settings, the materials through which radar waves pass may contain many small point targets that generate good reflections very small in size (e.g., the small ceramic artifacts in figure 3.16), which can only be described as *clutter* (if they are not the target of the survey). The amount of clutter visible in a reflection profile is dependent on the wavelength of the radar energy being propagated. If both the features to be resolved and the discontinuities producing the clutter are on the order of 25 percent of one wavelength, or a little less in maximum dimension, and about the same size, then reflection profiles may appear to contain only clutter, and there can be no discrimination between the two. Clutter can also be produced by large discontinuities, such as cobbles and boulders, but only when a lower-frequency antenna that produces a long wavelength is used.

There will always be radar energy that propagates in many complex orientations depending on frequency changes and the complexities of materials in the ground. To minimize the amount of reflection data derived from the sides of a survey line (called *side scatter*), the long axes of the antennas are usually aligned perpendicular to the survey transect. This allows the cone of transmission to be elongated in an in-line direction (figure 3.18). If there are narrow elongated features in the subsurface that are parallel to the direction of antenna travel (and therefore parallel to the electrical field generated by the antenna), only a small portion of the radar energy will be reflected back to the surface. In this case, if the antennas were not traversing almost directly on top of the buried linear feature, it might not be visible. Elongated buried features of this sort would usually have to be oriented perpendicular to direction of antenna travel in order to be visible on GPR profiles, and they would be visible as distinct "point sources" with noticeable reflection hyperbolas when crossed in this orientation (figure 3.15). Electrical engineers have developed a number of different antenna designs that produce other types of radiation patterns, most of which are not used in standard GPR surveys for archaeology (Annan and Cosway 1992). Some have also experimented with many different antenna radiation patterns and orientations along transects, which necessitate very different data processing and interpretation methods that are not discussed here (Jol and Bristow 2003).

RADAR PROPAGATION AND REFLECTION COMPLICATIONS
Ground Coupling

When a dipole antenna is placed on the ground, a major change in the radiation pattern occurs, owing to *ground coupling* (Engheta et al. 1982). This coupling occurs as the electromagnetic waves move from transmission in the air to transmission within the ground. During this process, refraction occurs as the radar energy passes through surface units, creating a change in the geometry of the propagating radar beam, with most of the energy channeled downward in a more focused cone from the transmitting antenna (Annan et al. 1975). The higher the RDP of the surface material through which the energy passes, the lower the velocity of the transmitted radar energy, and the more focused (less broad) the conical transmission pattern becomes (Goodman 1994). This focusing effect continues to occur as radar waves travel into the ground and materials of higher and higher RDP are encountered (figure 3.22). The amount of energy refraction that occurs with depth, and therefore the amount of focusing of the conical beam, is a function of Snell's Law (Sheriff 1984). According to Snell's Law, the amount of reflection or refraction that will occur at a boundary between two media depends on the angle of incidence and the velocity changes that occur at the interface. In general, the greater the increase in RDP with depth (which is the case in most ground conditions), the more focused the cone of transmission becomes (figure 3.22). This is because soil and sediment layers deeper in the ground are usually more compact and also have higher water saturations, both of which lead to higher RDP values. Radar beams could theoretically broaden the deeper energy

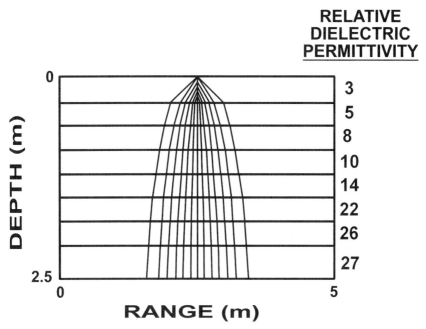

FIGURE 3.22
Energy Focusing with Depth. The radar energy cone of transmission will often become focused as energy travels in successive layers in the ground of increasing RDP, which is common for most field conditions.

moves in the ground if units of lower RDP were encountered, but this would be a rare phenomenon.

The type of surface materials within which the radar energy is coupled will also greatly affect the amplitude of the reflected waves below it. In figure 3.23, good radar reflections were recorded to about 40 nanoseconds through a surface material of limestone cobbles, but when the antennas crossed asphalt, the amount of energy coupled with the ground greatly increased. Reflections along the same transect were recorded at greater than 100 nanoseconds when the ground surface consisted of a good coupling material, illustrating how surface coupling can dramatically affect energy penetration depth and the resulting reflection amplitudes.

Radar energy coupling variations are also important because these changes in propagation depth and reflection amplitudes can be confused with "real" variations in reflectivity of materials in the ground when viewed in profiles.

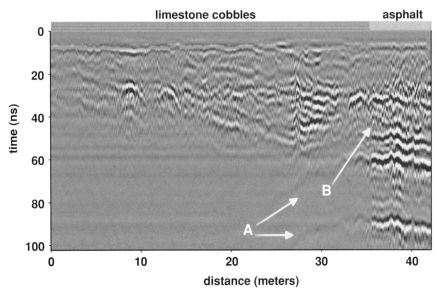

FIGURE 3.23
Coupling Changes Due to Differences in Surface Materials. Energy coupling with the ground can be variable, depending on the type of surface materials present. In this profile, collected along a street in southern Portugal, limestone cobbles had poor coupling properties, and energy propagated to only about 40 to 50 nanoseconds. Reflections that are visible at 60 and 90 nanoseconds on the right side of the reflection profile are barely visible (A) under the limestone cobble pavement. When the antennas crossed onto asphalt (B), coupling improved dramatically.

As antennas move over different surface materials (figure 3.24) or are moved over surface obstructions such as rocks or tufts of grass, coupling will also change, sometimes drastically and in a very short distance along a transect. This can create many strange and difficult-to-interpret reflection profiles with differing amplitudes of waves, none of which are reflecting "real" subsurface conditions. Uniform coupling, with little variation in the transmitted waveform along radar transects, occurs on paved surfaces and flat ground or areas with mowed grass. In uneven ground conditions, great care must be taken to keep the antennas close to the ground and at about the same orientation with the surface.

Frozen ground, especially with a thin layer of snow, can often be an excellent energy coupling surface producing little variation in transmitted waveform. In fact, some of the highest-quality GPR data have been acquired in frozen ground conditions, with excellent depth penetration and the recording of high-

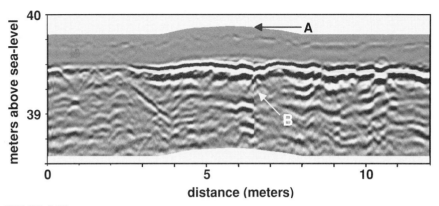

FIGURE 3.24
Coupling Changes Producing Anomalous Reflections. When antennas move over uneven ground and clumps of vegetation (A), antenna coupling changes the nature of waves traveling through the ground, producing anomalous amplitude changes (B), which can be misinterpreted as geological or archaeological changes.

amplitude waves. Data collected over the same ground after the ground thaws can often yield disappointing results, because of a drastic difference in energy coupling. One survey in particular that was conducted in Colorado is notable because the ground was frozen when data collection started in the early morning, producing high-quality reflection data. As the day progressed, and the sun slowly melted the ground surface, a noticeable decrease in the penetration depth of the energy was evident along with a decrease in amplitude of the reflected waves. This change was totally the result of differences in energy coupling and not differing constituents of materials in the ground.

Background Noise

An additional complication that affects resolution of reflections in the ground is *background noise*, which is almost always recorded during GPR surveys. Ground-penetrating radar antennas employ electromagnetic energy of frequencies that are similar to those used in television, FM radio, and other radio communication bands, so there are almost always nearby noise generators of some kind (figure 3.2). If there is an active radio transmitter in the vicinity of the survey, then there may be more interference than usual, but even when far away from the city, there will usually be background noise of some kind. Most radio transmitters have a very narrow bandwidth; if its frequency is known, it can be determined in advance, and an antenna frequency can be selected that is as far away as possible from any frequency that might generate spurious reflections in the data. With the wide bandwidth of most GPR antennas, however, it is usually difficult to completely avoid such external transmitter effects, and any major adjustments in antenna frequency may affect survey objectives.

External electromagnetic noise usually only becomes a significant problem if a study site is located in a city, near a military base, airport, or radio transmission antennas. One survey conducted near a U.S. military research site was occasionally disrupted by radar noise from unknown sources. This noise was visible during collection as periodic very high-amplitude reflection traces that totally overwhelmed any reflections derived from within the ground (figure 3.25). Many different frequency-filtering schemes were attempted, to no avail.

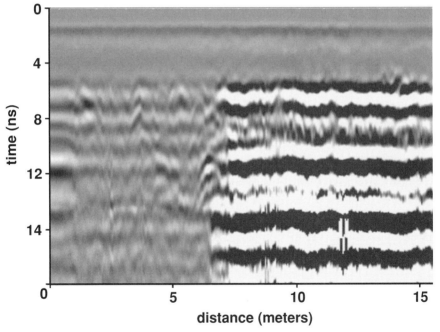

FIGURE 3.25
Extreme Electromagnetic Noise in a Reflection Profile. Sometimes electromagnetic energy noise generated from a source nearby will totally overwhelm energy recorded at a surface antenna from within the ground. This reflection profile shows the onset of a "blast" of noise producing very high-amplitude reflections from 7 to 16 meters along the transect. The source of this noise, near the Los Alamos National Laboratory, in New Mexico, was never discovered and lasted about 15 minutes.

Finally, in about an hour, the noise completely disappeared, and data collection could continue. The source of the interfering electromagnetic energy was never discovered.

The recent proliferation of cellular telephones that can be in use nearby during data acquisition has also become a problem, as they produce FM radio noise in the same frequencies as some GPR antennas. This type of noise has been observed when collecting GPR data near a busy road where interference generated by passing motorists using cellular phones periodically disrupted recorded reflections.

Focusing and Scattering Effects

Reflection off a buried surface that contains ridges or troughs, or any other irregular features, can either focus or scatter radar energy, depending on the surface's orientation and the location of the antenna on the ground surface. If a reflective subsurface plane is slanted away from the surface antenna's location or is shaped so that the surface is convex upward, most energy will be reflected away from the antenna, and no returning energy, or a very low-amplitude reflection, will be recorded (figures 3.21 and 3.26). This is termed radar *scatter*. The opposite is true when the buried surface is tipping toward the antenna or the surface is concave upward (figure 3.26). Reflected energy in this case will be focused, and a very high-amplitude reflection derived from a portion of the buried surface would be recorded.

Figure 3.26 illustrates an archaeological example of the focusing and scattering effects when a narrow buried moat is bounded on one side by a trough and the other side by a mound. When the radar antenna is located to the left of the deep moat, some of the reflections are directed back to the surface antenna, but there is some minor scattering, generating a weaker reflection than usual from the buried surface. When the antennas are located directly over the deep feature, there will be a high degree of scattering, and much of the radar energy, especially that which is reflected off the sides of the moat, will be directed away from the surface antenna and not be recorded. This scattering effect would make the narrow moat almost invisible in reflection profiles. When the antenna is located directly over the wider depression to the right of the moat, there will be focusing of the radar energy, creating a higher-amplitude reflection from this portion of the subsurface interface.

This focusing and scattering condition is quite common, and it can occur often repeatedly along one buried planar surface. It was noticed when mapping

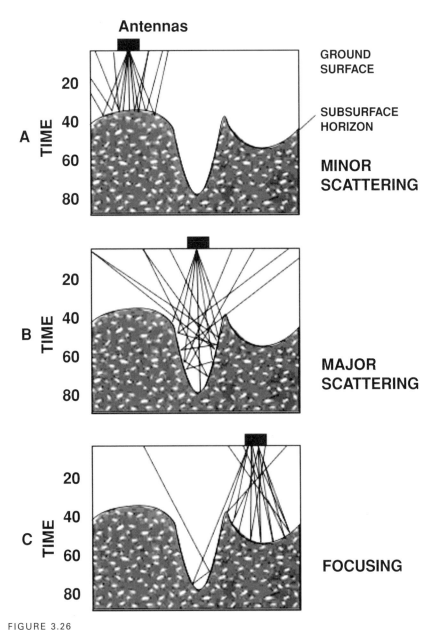

FIGURE 3.26
Energy Scattering and Focusing. Radar energy transmitted from a surface antenna will be scattered from a convex upward interface (A). Deep trenches or other near-vertical features (B) will produce a great deal of scattering, and little energy will reach the surface antenna to be recorded, making features of this sort almost invisible in reflection profiles. Concave upward features (C) will focus radar energy, producing high-amplitude reflections.

the surface of a buried prehistoric agricultural field, consisting of uniform ridges and furrows (figure 3.27). The furrows on this buried interface focused the radar energy, creating higher-amplitude reflections while the ridges scattered it, producing only weak reflections or none at all. The undulating surface itself was not discernible in reflection profiles, as a 300-megahertz antenna was used, which tended to blur somewhat the reflection from the buried interface. It was the periodic changes in the wave amplitudes that were a hint that these types of buried features were present. The layer was later excavated, and the geometry of the bedding plane was confirmed to be a buried ancient agricultural field (figure 3.28).

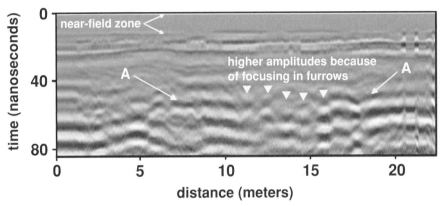

FIGURE 3.27
Scattering and Focusing on a Horizontal Reflection Surface. Varying high and low amplitude reflections along reflection surface (A) were found to be small ridges and furrows in a buried agricultural field at the Ceren site in El Salvador.

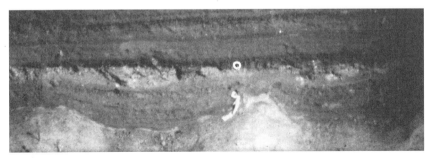

FIGURE 3.28
Buried Agricultural Field That Causes Focusing and Scattering. Excavation of the reflecting surface A, shown in figure 3.27, shows the ridges and furrows of a prehistoric agricultural field, which alternately focused and scattered reflected radar waves.

The Near-Field Effect

Energy radiated from a surface antenna generates a strong electromagnetic field around the antenna within a radius of about 1.5 wavelengths of the center frequency (Balanis 1989; Engheta et al. 1982; Kraus 1950; Sheriff 1984). Within this zone, coupling of the radar energy is occurring with the ground, generating an advancing wave front of propagating waves in the standard conical transmission pattern. It can be said that the ground within about 1.5 wavelengths of a standard dipole antenna is technically "part of the antenna" in that no radiation is occurring within this zone and therefore technically no wave propagation. This *near-field zone* is usually visible in GPR profiles as a region of little or few reflections beginning at the ground surface and continuing to some depth (figure 3.21). In the GPR literature, this zone is sometimes incorrectly called the near-surface zone of interference. For the 10-, 100-, and 1,000-megahertz antennas, the near-field zones are approximately 30 meters, 3 meters, and 30 centimeters, respectively, but vary depending on the downloaded frequency.

If low-frequency antennas are used, the near-field zone where few significant reflections are generated can sometimes be between 2.5 and 5 meters of the ground surface. If the target features are located within the near-field zone, it is unlikely that they will be visible in GPR profiles, and a higher-frequency antenna should be used. There can, however, sometimes be important reflection data recorded within the near-field zone, even if reflections are not immediately visible on standard two-dimensional reflection profiles. Due to the wide bandwidth of radar transmission, some high-frequency (shorter-wavelength) energy will still be generated even from a lower-frequency antenna, which will couple with the ground at a much shallower depth, and some shallow reflections can still be generated and visible in profiles within what is broadly defined as the near-field zone. If these reflections are high enough in amplitude, they will still appear as weak reflections within the otherwise reflection-free near-surface layer. Some subtle reflections in the near-field may never be noticeable in standard two-dimensional profiles but can become visible after the data are computer processed to produce amplitude slice maps, which are discussed in chapter 7. Other very weak, but important, reflections in the near-field can also be enhanced in profiles by increasing amplitudes in the shallowest portion of the profiles with the aid of computer software. This technique, called *range gaining*, is usually performed during equipment setup prior to collecting data (chapter 4), but it can also be applied after returning from the field, which is discussed in chapter 6.

Air Waves and Near-Surface Obstructions

The wide field of energy transmission from most GPR antennas can produce unwanted reflections that occur from features that may not be in the ground, especially with lower-frequency antennas that are not well shielded. When using unshielded antennas in areas of high-tension power lines or nearby buildings, reflections are likely to be produced from these features, creating what are called *air waves* in reflection profiles (figure 3.29). These reflections are often high in amplitude and can obscure meaningful reflections from within the ground. They are called air waves because the radar energy that produces them travels to and from the unshielded GPR antennas in the air. As the velocity of radar transmission in air does not change like it does in the ground, air waves produce very straight reflections that occur at one time in the profiles when antennas are moved parallel to the reflecting surface. When antennas are moved toward or

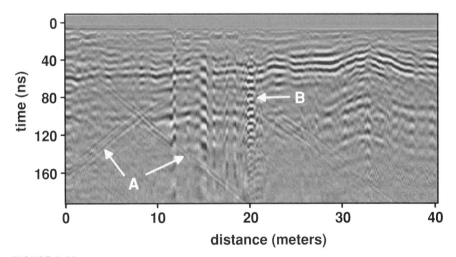

FIGURE 3.29
Air Waves. Radar energy that travels in air from the antenna, to an object on the surface, and back to the antenna is recorded as air waves (A), visible as sloping, almost straight reflections. This reflection profile was collected with unshielded 300-MHz antennas along the edge of a road in El Salvador where nearby trucks reflected energy. Three pieces of buried metal can also be seen as multiple horizontal reflections, one of which is noted as (B).

away from a surface reflecting source, the energy traveling in air will be recorded as a straight reflection that gradually increases or decreases in time (figure 3.29). A GPR survey was once attempted with poorly shielded 300-megahertz antennas in a parking lot, bounded on three sides by tall buildings, and there were many more air wave reflections recorded from the buildings than from within the ground, making the survey results essentially unusable.

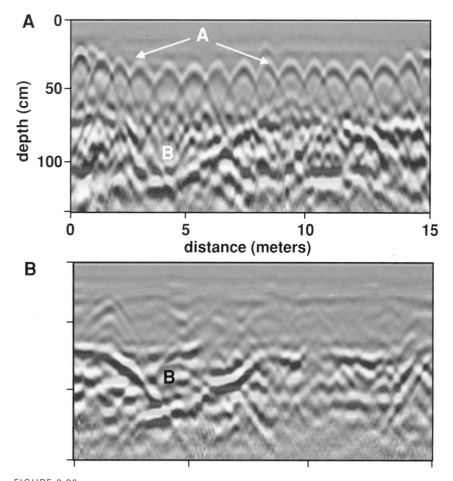

FIGURE 3.30
Near-Surface Metal Interference. Metal reinforcing bars buried in surface concrete produce evenly spaced hyperbolic reflections (A) in profile A collected over a buried trench (B). Enough energy is transmitted through the metal obstructions to produce a useful image of the buried features below. In profile B, located 50 centimeters away and parallel to profile A, the surface pavement contained no metal, and the trench (B) is much better defined. These profiles were collected in downtown Reno, Nevada.

The air wave problem was encountered to a large extent within a deep excavation whose sides had been supported by metal barriers to keep them from collapsing (Carrozzo et al. 2003). The high-amplitude air waves generated from the metal barriers either overwhelmed or interfered with reflections from within the ground, necessitating a complex series of processing steps to remove them. This was done by first producing a mean profile of the air wave reflections generated from the metal walls, which was possible because the walls were a known distance away from the antennas, and the velocity of the air waves was known (the speed of light) and their arrivals were calculated. The mean reflection profiles of the predicted air waves were then subtracted from the actual reflection profiles to produce residual reflection profiles, which were then processed to increase the remaining amplitudes generated from within the ground. Good results were achieved, but only after significant and lengthy data-processing steps.

Surface or near-surface metal objects can often create multiple reflections that "ring down" through a profile or create a number of shallow reflection hyperbolas, obstructing some or all features below them (figure 3.30). This is very common when conducting GPR surveys on road surfaces where metal reinforcing bars are used to stabilize concrete or where there are an abundance of buried pipes. Buried metal wire or bars used for reinforcing in roadways and sidewalks will often produce many equally spaced hyperbolic reflections, which are very distinct in reflection profiles (figure 3.30). Even with this kind of near-surface energy obstruction, some radar energy will still travel between the metal, reflecting from other materials buried deeper in the ground, and usable data can still sometimes be obtained.

4
Ground-Penetrating Radar Equipment and Acquisition Software

GPR SYSTEMS

A number of different manufacturers of GPR units are typically used in archaeological surveys. The most commonly used units in North America are manufactured by Geophysical Survey Systems Incorporated (GSSI), located in North Salem, New Hampshire, and Sensors and Software Inc. of Mississauga, Ontario, Canada. GPR manufacturers in Europe and Japan also market multiuse GPR systems with excellent archaeological applications. Most of the systems produced for general-purpose GPR surveys employ antennas that transmit pulsed radar energy of one center frequency. Sensors and Software antennas were originally developed to be moved in steps, recording data at specified intervals along transects in grids, but recent models can operate in both continuous and survey wheel acquisition mode.

All GPR systems on the market have the ability to collect data along transects using some kind of survey wheel that measures distance and spaces reflection traces equally along transects (figure 3.5). Some survey wheels roll along the ground surface, and their revolutions and the number of reflection traces collected are calibrated for distance. For uneven ground, a small "hip wheel" has been developed that can attach to one's belt or the antennas, which feeds out thread as the antennas are moved along a survey transect. In this collection method, a small wheel in the device contains a large spool of thread, and the wheel's revolution is calibrated for distance. A certain number of turns of the wheel determines the number of reflection traces to collect per distance of ground covered as the thread is spooled out. With this method, at the beginning of each transect, the thread is tied to a stake in the ground that is spooled out as the antennas are moved. When each transect is recorded, the thread is discarded; at the end of a large survey, strands of thread litter the survey area.

There are many similarities and some differences between models and manufacturers of GPR units. Many newer models are compact enough so that one person can theoretically collect large numbers of transects alone. These systems place the antennas, power source, and control system on a cart that can be rolled along the ground, or placed in a backpack, relieving the antenna operator from having to be tethered to a base station (figure 3.7). Most can operate for many hours on rechargeable lightweight batteries, while others run on 12-volt batteries or household electric currents. All GPR systems contain one or more internal hard drives that store a software program to drive data acquisition and for reflection data storage. Systems typically have a number of different ports that allow acquired data to be transferred to a personal computer after a survey is complete. Newer models save data on flash memory cards or chips that can hold many hundreds of megabytes of data with little power usage, making these systems very lightweight and allowing data to be quickly downloaded to other computers for processing.

All GPR systems employ a computer screen that is necessary for presurvey calibration and allows reflection profiles to be viewed in real time during collection. Screens are sometimes built into the control box of the system or one can be used from a peripheral laptop computer (figure 2.2).

All manufacturers are continually introducing new GPR system models, striving for compactness, longer-life battery usage, transportability, and ease of data collection. There have been some recently developed systems that are "dumbed-down" versions of more standard GPR units, which can be used to find pipes or voids in the ground quickly for buried utility location that don't require complex postacquisition processing. These simple systems have little ability to be calibrated to unusual field conditions or store complex data sets that can later be processed with GPR software and should be avoided for most archaeological applications.

There is one type of GPR system that uses a "stepped-frequency" technology, which focuses a narrow radar beam of varying frequencies into the ground. This technique, which uses a coiled antenna system (Noon et al. 1994; Tomizawa et al. 2000; Valle et al. 2000), has not yet seen significant archaeological application and will not be discussed further in this book. Others can be used to collect radar reflections within or between bore holes (Wright and Lane 1998) or by placing a transmitting antenna in a hole below a known archaeological feature and collecting radar pulses from an array of antennas located on the ground surface above.

Standard GPR systems consist of three main elements, the control unit (pulse generator, computer, and associated software), the antennas (paired transmitting and receiving antennas), and the display unit (computer screen) (figure 2.2). The control unit produces a high-voltage electrical pulse, which is sent through cables to the transmitting antenna, which amplifies the voltage and shapes the pulse, emitting it. Cables that transmit the electrical pulses come in varying lengths and are manufactured of either coaxial copper filament or fiber optic material. These cables connect the antennas to the system that might be located at a base station or on a survey cart or backpack. Fiber optic cables, which transmit a digital signal to and from the antennas, greatly reduce some of the equipment-related noise that can affect the signal clarity in some coaxial cables (Davis and Annan 1989). Fiber optic connectors, however, are not very "field durable," and the smallest amount of dust on a connection, or any rough wear during data acquisition, can cause them to malfunction. When only short fiber optic cables are needed to connect peripheral devices with the control unit, and all components are being transported on a cart or backpack, wear and tear of fiber optic cables can be minimized.

All GPR systems generate the electrical pulse necessary to create propagating radar waves directly at the antenna. Some systems transmit the waveform of the received wave returning from within the ground to the control system for recording as digital data, or as analog data in the form of voltage changes, which can then be digitized and stored at the control system (Annan and Davis 1992). Better data quality is usually obtained when the transmitted and received waves travel the least distance in analog cables, where system noise can be generated.

Most GPR systems use a hand-held marker device to record the surface position (fiducial mark) of the antennas along a survey transect during reflection data acquisition. When the marker is activated, some early-model control units impose a noticeable high-frequency sine wave, covering one recorded reflection trace in order to define the mark locations. When digital data are acquired, a fiducial mark is usually inserted into the data string as an identifiable bit of data in each reflection trace where a mark is placed. During data collection, the marker button can either be located on the antenna and operated by the person pulling it or attached to the control unit and operated by someone, at the command of the antenna operator. In continuous data acquisition, the marker button must be pushed at standard intervals along the antenna transects, usually every few meters or less, at stations that have been presurveyed using a tape measure. If data are collected with a survey wheel, and specific distances along transects do not need to be recorded in the data string, fiducial marks can be used to identify the location of surface obstacles or other features of note. When using the step method of data acquisition, the antennas' location on the ground is predetermined by the step spacing, and no marker device is necessary for navigation purposes.

DATA ACQUISITION SOFTWARE: SETUP PARAMETERS

Manual adjustments are always necessary prior to conducting any GPR survey (Kemerait 1994). In newer GPR units, some of the equipment adjustments can be automatically controlled by the acquisition software, but many of these can be (and usually should be) manually overridden. In older models, most of the pre-collection adjustments needed to be made manually with switches or knobs located on the control unit. Modern units use a software interface to do this, which can be controlled from a keyboard or touch pad or using an attached laptop computer.

Header Information

Most digital GPR units have a procedure where general header information can be input for each file or the grid as a whole. This information typically includes the date of the fieldwork, antenna frequency, site name, grid name or number, and other pertinent information or comments. Many units allow most of this information to be entered at the start of acquisition, and it can be modified for each transect and file within a grid, if desired. Often many of the significant acquisition settings discussed here are automatically recorded in the headers of each file or sometimes a separate file altogether for each profile, and they can be viewed later on, if good field notes are not kept, or lost.

Each profile within a grid is usually automatically saved as a separate file to the computer hard drive or other media as they are collected. The names and sequence of these data files should also be noted in a field book as they are being recorded. Files are usually recorded sequentially, with the first reflection profile in a grid saved as file1, and so on, as they are collected. To avoid confusion when more than one grid of data are being collected in a day, it is usually good to start each profile in a grid as file1 in a separate computer directory, and always keep good notes of their locations and orientations within those grids.

Time Window

All GPR systems allow the user to select the time period over which reflection data are recorded. The *time window* is defined as the amount of two-way travel time, measured in nanoseconds, that the receiving antenna will "listen" and record the reflected radar wave energy (figure 3.6). This window will normally open just before the radar pulse is transmitted and is closed after all reflections of interest, from the depth desired in the ground, have been recorded. If the velocity of the material and the approximate depth of the features to be resolved are known, the amount of time necessary for radar energy to travel down to and then be reflected back from the zones of interest can be estimated. The time window can then be adjusted so that it is open for at least this period so that all important reflections in all antenna transects within the survey grid are recorded. It should usually be adjusted so that more reflection data, from a greater depth, are being recorded than is necessary. Often, due to unforeseen subsurface velocity changes, reflections from features of interest could possibly be received at times later than preliminary calculations estimate, and if the time window is not open for long enough, they will not be recorded. It is also possible that buried horizons of interest might dip to greater depths or be covered with a greater thickness of overburden in some portions of a grid than initial estimates, also necessitating a longer time window in order to record them.

In most archaeological applications, a time window of 100 nanoseconds (two-way travel time) or less is usually sufficient to record reflections within 2 to 3 meters of the surface, depending on the velocity of radar wave propagation. In a material with a relative dielectric permittivity of 8, a 20-nanosecond window is capable of recording reflections to about 1 meter depth (table 4.1). These types of depth calculations are always independent of antenna frequency.

Table 4.1. Depth in Meters to a Reflector Through Media of a Given Relative Dielectric Permittivity

(ns)	Relative Dielectric Permittivity															
	1	2	3	4	5	6	7	8	9	10	15	20	30	40	50	60	80
10	1.50	1.06	0.87	0.75	0.67	0.61	0.57	0.53	0.50	0.47	0.39	0.34	0.27	0.24	0.21	0.19	0.17
20	3.00	2.12	1.73	1.50	1.34	1.22	1.13	1.06	1.00	0.95	0.77	0.67	0.55	0.47	0.42	0.39	0.34
30	4.50	3.18	2.60	2.25	2.01	1.84	1.70	1.59	1.50	1.42	1.16	1.01	0.82	0.71	0.64	0.58	0.50
40	6.00	4.24	3.46	3.00	2.68	2.45	2.27	2.12	2.00	1.90	1.55	1.34	1.09	0.95	0.85	0.77	0.67
50	7.50	5.30	4.33	3.75	3.35	3.06	2.83	2.65	2.50	2.37	1.94	1.68	1.37	1.19	1.06	0.97	0.84
60	8.99	6.36	5.19	4.50	4.02	3.67	3.40	3.18	3.00	2.84	2.32	2.01	1.64	1.42	1.27	1.16	1.01
70	10.49	7.42	6.06	5.25	4.69	4.28	3.97	3.71	3.50	3.32	2.71	2.35	1.92	1.66	1.48	1.35	1.17
80	11.99	8.48	6.92	6.00	5.36	4.90	4.53	4.24	4.00	3.79	3.10	2.68	2.19	1.90	1.70	1.55	1.34
90	13.49	9.54	7.79	6.75	6.03	5.51	5.10	4.77	4.50	4.27	3.48	3.02	2.46	2.13	1.91	1.74	1.51
100	14.99	10.60	8.65	7.50	6.70	6.12	5.67	5.30	5.00	4.74	3.87	3.35	2.74	2.37	2.12	1.94	1.68
110	16.49	11.66	9.52	8.24	7.37	6.73	6.23	5.83	5.50	5.21	4.26	3.69	3.01	2.61	2.33	2.13	1.84
120	17.99	12.72	10.39	8.99	8.04	7.34	6.80	6.36	6.00	5.69	4.64	4.02	3.28	2.84	2.54	2.32	2.01
130	19.49	13.78	11.25	9.74	8.71	7.96	7.37	6.89	6.50	6.16	5.03	4.36	3.56	3.08	2.76	2.52	2.18
140	20.99	14.84	12.12	10.49	9.39	8.57	7.93	7.42	7.00	6.64	5.42	4.69	3.83	3.32	2.97	2.71	2.35
150	22.48	15.90	12.98	11.24	10.06	9.18	8.50	7.95	7.50	7.11	5.81	5.03	4.11	3.56	3.18	2.90	2.51
160	23.98	16.96	13.85	11.99	10.73	9.79	9.07	8.48	7.99	7.58	6.19	5.36	4.38	3.79	3.39	3.10	2.68
170	25.48	18.02	14.71	12.74	11.40	10.40	9.63	9.01	8.49	8.06	6.58	5.70	4.65	4.03	3.60	3.29	2.85
180	26.98	19.08	15.58	13.49	12.07	11.02	10.20	9.54	8.99	8.53	6.97	6.03	4.93	4.27	3.82	3.48	3.02
190	28.48	20.14	16.44	14.24	12.74	11.63	10.76	10.07	9.49	9.01	7.35	6.37	5.20	4.50	4.03	3.68	3.18
200	29.98	21.20	17.31	14.99	13.41	12.24	11.33	10.60	9.99	9.48	7.74	6.70	5.47	4.74	4.24	3.87	3.35
210	31.48	22.26	18.17	15.74	14.08	12.85	11.90	11.13	10.49	9.95	8.13	7.04	5.75	4.98	4.45	4.06	3.52
220	32.98	23.32	19.04	16.49	14.75	13.46	12.46	11.66	10.99	10.43	8.51	7.37	6.02	5.21	4.66	4.26	3.69
230	34.48	24.38	19.91	17.24	15.42	14.08	13.03	12.19	11.49	10.90	8.90	7.71	6.29	5.45	4.88	4.45	3.85
240	35.98	25.44	20.77	17.99	16.09	14.69	13.60	12.72	11.99	11.38	9.29	8.04	6.57	5.69	5.09	4.64	4.02
250	37.48	26.50	21.64	18.74	16.76	15.30	14.16	13.25	12.49	11.85	9.68	8.38	6.84	5.93	5.30	4.84	4.19
260	38.97	27.56	22.50	19.49	17.43	15.91	14.73	13.78	12.99	12.32	10.06	8.71	7.12	6.16	5.51	5.03	4.36
270	40.47	28.62	23.37	20.24	18.10	16.52	15.30	14.31	13.49	12.80	10.45	9.05	7.39	6.40	5.72	5.23	4.53
280	41.97	29.68	24.23	20.99	18.77	17.13	15.86	14.84	13.99	13.27	10.84	9.39	7.66	6.64	5.94	5.42	4.69
290	43.47	30.74	25.10	21.74	19.44	17.75	16.43	15.37	14.49	13.75	11.22	9.72	7.94	6.87	6.15	5.61	4.86
300	44.97	31.80	25.96	22.48	20.11	18.36	17.00	15.90	14.99	14.22	11.61	10.06	8.21	7.11	6.36	5.81	5.03

Determining the optimum time window in advance of data collection is extremely important. Some materials may have very high relative dielectric permittivities, and therefore radar energy travels through them at very slow rates. If that material also had a low conductivity, energy could conceivably travel quite deep in the ground but at a slow rate, and one would need to collect reflections over a longer time window. For instance, if an average surface soil is assumed with an RDP of 9, a 30-nanosecond time window would allow the collection of reflections to about 1.5 meters in the ground (see table 4.1). If later on it was determined that the actual RDP of the soil was 20 (perhaps it had a good deal of water in it that was not accounted for), then the radar propagation velocity would be much slower and that same time window would only have allowed collection to about 1 meter in the ground (again, see table 4.1). If this were determined only after data had been collected, and the features of interest were located between 1 and 1.5 meters in the ground, the whole survey would have to be repeated as the pertinent reflections from the depth of interest would not be within the programmed time window. This sad scenario has happened more often than many GPR practitioners would like to admit. It can only be overcome by spending a great deal of attention to local conditions, doing velocity analysis in advance (chapter 5), and adjusting setup parameters for the correct time window prior to data collection.

Samples per Reflection Trace

Once the time window is set, the number of samples necessary to record a reflected waveform must be selected for all digital GPR units. One sample is a digital value that defines a portion of the reflected waveform. The more digital samples there are to define a wave, the more accurate the form of that reflected wave becomes. The longer the time window is open, the larger number of samples are usually necessary to adequately define the reflection trace of the reflected wave.

Any number of data samples can be selected to define each reflection trace on most units, but by convention it is usual to select 512; however, 1,024 and 2,048 are also common sampling rates. Due to incremental sampling, which digitize one bit of data for each pulse generated, if 512 samples are selected to define each reflection trace, then there must also be 512 pulses transmitted into the ground in succession in order to record each trace. If this were the case, and a large number of reflection traces were also programmed to be collected every second, some GPR systems would not be capable of generating enough pulses and recording enough samples per second to adequately record the data programmed into the system.

The maximum resolution (defined by the shape of the waveform) that can be obtained is also dependent on the wavelength of the reflected waves that are generated by the antenna, which is a function of the antenna frequency (table 3.2). Higher-frequency antennas generate shorter-wavelength waves that are recorded in quick succession, which may need more digital samples to define them within a given time window, as they have a very complex waveform. When all these factors are considered, it is easy to see that a number of estimates and assumptions are necessary prior to determining the sampling definition. Some experimentation may be necessary in the field while the antennas are stationary in order to obtain the optimum sampling rate for the wavelength produced and time window allotted.

It is just as important in preacquisition adjustments to make sure that the time window is not open for too long as it is for too short a time. The longer the time window is open, the more samples per reflection are necessary for good resolution of the recorded waveform, and the more samples are necessary to record. If too much reflection data are being collected over a long time window beyond the depth of interest, too many samples would be needed to define the waves from depths at the far end of the range. Most of the digital data recorded would then be defining reflections far outside the depth of interest. If this were the case, storage capacity on a hard drive or other media could fill up quite quickly with useless data, especially if a large survey is being conducted.

Trace Stacking

The *stacking* of reflection traces, done with a *horizontal filter*, or *spatial filter* (sometimes termed *horizontal smoothing*), can be applied prior to data acquisition or later during postacquisition processing (discussed in chapter 6). This data manipulation method arithmetically averages digital values of successive reflection traces so that one composite trace is recorded every certain distance along a transect (Fisher et al. 1992; Grasmueck 1994; Maijala 1992). Many GPR systems allow the operator to manually adjust the stack rate to any integer. This process will average sequentially any number of programmed traces, recording only the average from each of these sequentially, usually as a running average (Davis and Annan 1989). It is important to recognize that the more reflection traces that are stacked, the slower the antenna must be moved along the ground surface in continuous collection mode, or the more traces that must be recorded every distance along a transect in survey wheel mode, in order to achieve the same number of composite reflection traces per unit of ground covered.

This type of horizontal filtering of sequential traces will effectively remove waveforms that may have been generated from surface irregularities such as small bumps or dips in the ground surface (Fisher et al. 1992; Grasmueck 1994; Maijala 1992). It also filters out the effects of velocity changes due to minor water saturation changes, small rocks or voids in the subsurface, and changes in amplitude due to antenna coupling differences with the ground.

Stacking is usually a good idea when the antennas are collecting in continuous mode and are moving at a fairly slow speed (at an average human pace or less), and when fairly large features in the ground are the target. A trace stack of eight or more contiguous traces into each one that is recorded can then improve the quality of the subsurface reflection data and still yield good subsurface coverage. In this fashion, if 64 reflection traces are being measured every second in continuous data collection mode, and each 8 consecutive traces are stacked into one composite, then 8 would be recorded each second. If the radar antennas are moving at a rate of 16 centimeters per second along the ground surface, then there will be one reflection trace recorded for every 2 centimeters of ground covered (16 divided by 8). If a stack of 16 were applied (each 16 consecutive traces averaged into one), then the subsurface coverage would be one recorded reflection trace every 4 centimeters along the same transect.

The same method can be used to stack traces in survey wheel or step acquisition mode in order to average out reflections and smooth the reflection data. Unless the ground surface is very uneven or there is highly variable stratigraphy in the subsurface (that is not of importance to the questions at hand), it is usually best to collect all the reflection traces available and stack them later during data processing, if it is later determined to be necessary.

Transmission Rate

Radar systems typically transmit at a rate of more than 50,000 pulses per second (often measured in kilohertz [KHz]). With the presently available GPR technology, it is impossible to record each individual reflected trace generated from each transmitted pulse due to the rapidity at which the pulses are being transmitted, and then reflected back to the surface. To overcome this problem, radar systems sample incrementally, meaning that one pulse must be transmitted for each sample that is recorded. If the system were set up to stack 16 sequential traces into one recorded reflection trace, then there must be at least 512 pulses times 16 (8,192) transmitted for every one reflection trace recorded.

An understanding of transmission and recording rate usually becomes important only when determining the horizontal resolution of the recorded reflection data. Depending on the speed at which the antennas are moving across the ground, adjustments of the recording and stacking rates may be necessary in order to obtain good subsurface coverage. For instance, if it is necessary to record 4 complete reflection traces (defined by 512 samples) per second (after stacking 16 traces into one), then 8,192 sequential pulses times 4 (32,768) must be transmitted each second. If the radar system being used is only capable of transmitting at a rate of 25,000 pulses per second, there would not be enough pulses transmitted to allow for 4 recorded reflection traces per second. If this were the case, a few minor adjustments would need to be made prior to recording data: (1) The stacking rate could be lowered; (2) the antennas would need to be moved over the ground slower, recording more total reflection traces per distance covered; or (3) the time window can be shortened, necessitating fewer samples to define each trace (or all three of the above).

If the time window is fairly short, the stacking is minimized, and if the sampling rate is kept at 512 samples per reflection trace or less, there are usually more than enough pulses being transmitted into the ground in order to record the desired traces. The adjustments just discussed are usually required only if the antennas are moving at a high rate of speed (perhaps towed behind a vehicle), there are very high stack rates applied (more than 16 traces stacked into one), or an extremely high waveform resolution (many samples per reflection trace) is necessary in order to define the subsurface reflections within a large time window. These potential problems are partially overcome in many recently manufactured GPR units that transmit at rates much higher than 50,000 pulses per second, allowing much more latitude in these adjustments.

During survey wheel collection, if the sampling rate is not set high enough to allow the programmed number of reflection traces to be recorded per unit distance covered, most systems will usually notify the operator with a beep warning the operator to slow the antenna transport along the ground. This will then allow the sampling rate to "catch up" with the number of traces being recorded. When collecting data using the step method, there are always more than enough pulses being transmitted to allow for the recording of one reflection trace and the horizontal resolution becomes only a function of the distance between steps.

Time Zero Position

Prior to acquiring reflection data from within the ground a calibration must be made so that the first reflection recorded from any pulse emitted from the antenna is the reflection from the ground surface (also called the *direct wave*). This is done while the antenna is stable on the ground in the configuration that will be used for all subsequent data acquisition. When calibrated for the zero position, the first reflection visible is usually from the ground surface, and all subsequent reflections recorded in time will be received from horizons deeper in the ground (Yelf 2004). In all GPR systems with a video monitor, the direct wave can be displayed and is visible as the first large-amplitude wave after a period of no data recording (figure 3.6). Most GPR systems have an automatic programming procedure that will identify this first reflection and set the zero position so that the ground surface reflection is about 1 nanosecond or so into the time window, and all later reflections from deeper in the ground are recorded later in time. Older systems display the reflection trace on an oscilloscope in a similar fashion, and the first reflection from the ground surface must be manually adjusted, and then time zero can be set.

The time window should also always be placed so that the first recorded reflection from the ground surface is not at exactly time zero but just lagged a little below it, so that the ground surface can always be found in reflection profiles after returning from the field, if there is any question. Most systems will then lock that zero position, and all reflection profiles will be collected with the ground surface in the same place within the time window. Any time lag between zero time and the ground surface reflection can later be compensated for during data processing. If the automatic time zero function is not locked in position, it is possible that the computer that controls data acquisition might be constantly readjusting time zero as data are collected, and all traces in the ground will be recorded at different times relative to zero, making for a very confusing data set and potentially disastrous survey results.

Range Gains

Due to the conical spreading of the transmitted radar waves and the attenuation of radar energy as it passes through the ground, later reflection arrivals generated from deeper in the ground will almost always have lower amplitudes than earlier arrivals (figures 3.3 and 3.6). To recover these lower-amplitude waves, gain control (*range gaining*) is applied to all reflection traces in a profile during either acquisition or postacquisition processing (Jol and Bristow 2003; Leckebusch 2003; Maijala 1992; Shih and Doolittle 1984; Sternberg and McGill 1995). This will amplify those waves received from deeper in the ground so they are visible (figure 4.1). Range gain settings are standard on most GPR equipment, and most systems have software that will automatically adjust waveform amplitudes so they are all "on scale." Older systems must be manually adjusted (Fisher et al. 1994; Geophysical Survey Systems, Inc. 1987). There is usually a linear or exponential relationship between the amount of gain that must be applied to recorded wave amplitudes and the time it is received, with higher gains applied to reflections recorded later in the time window (figure 4.1).

There are two schools of thought with respect to range gains. One group of GPR practitioners and some GPR system manufacturers believe that no gains should be applied during data collection, and that amplitudes of reflected waves should only be adjusted after returning from the field. In this way, a raw data set will be collected that can be adjusted in any manner deemed appropriate after all reflection profiles are recorded within a grid. The other school of thought believes that gains should be applied at the time of acquisition so that the variables affecting amplitudes in the ground can be adjusted for immediately. In that way,

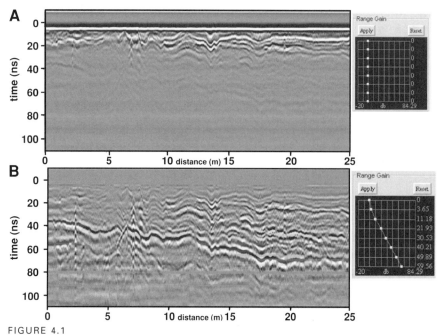

FIGURE 4.1
Range Gaining. Ungained reflection profiles in A have good near-surface resolution, but reflections recorded from deeper in the ground are low in amplitude and barely visible. When range gains are applied to the same reflection data, arithmetically enhancing the amplitudes from deeper in the ground, reflections become visible (B). Profile B shows a high-amplitude reflection in 300 MHz data, between 40 and 60 nanoseconds, generated from a sloping buried living surface at the Ceren site, El Salvador.

changes in soil moisture, the various depths of layers, and changing ground surface materials can be accounted and adjusted for. The idea in this latter method is that even if the gains were not perfectly applied in the field, postacquisition processing can be used to increase or decrease the amplitudes of reflected waves later on. But even this method of gaining can be fraught with problems. If the gains are set too high prior to collecting reflection profiles (increasing the amplitudes by large factors), and the antennas were then moved over an area with very reflective buried materials, then the increase in recorded amplitudes would "go off scale," and the highest values would be "clipped" and not recorded (figure 4.2). A recent "compromise" by one GPR manufacturer allows the user to set the gains in the field, but by default records amplitudes at 25 percent of the field settings to prevent inexperienced operators from amplitude clipping. This "dumbing down" of the acquisition procedures (which is becoming more common as GPR systems reach a broader number of users) is regrettable because one will still have to reprocess all reflection data anyway after returning from the field in an attempt to replicate the original field settings.

If one decides to set range gains in the field, it is very important to move the antennas over much of the ground to be surveyed during the calibration procedure before any reflection data are collected so that a general idea of the reflectivity (and therefore recorded wave amplitudes) of buried materials can be determined. In this way, the gains can be set at the location where the highest reflection amplitudes will likely be recorded and one can be fairly well assured that the remainder of the amplitudes recorded in all the reflection profiles within the grid to be surveyed will be "on scale" and not lost due to overgaining, resulting in amplitude clipping.

If the buried features of interest reflect little radar energy and therefore would be recorded as very low-amplitude reflections, it might be advisable to set the gains very high in order to be able to see them in the reflection profiles. This is done at the risk of clipping amplitudes of other features that might not be of interest. One survey was conducted in an area where large piles of buried brick rubble from collapsed walls were creating very high-amplitude reflections. The targets of the survey, however, were very subtle burial features in nearby graves adjacent to the collapsed brick walls. It was therefore decided to increase the gains of all reflections at all depths, which increased the amplitudes of the reflections from the bricks off scale (clipping them) but also increased the amplitudes of the subtle grave features so they were visible. The downside of this acquisition setup procedure was that the portions of the profiles crossing the

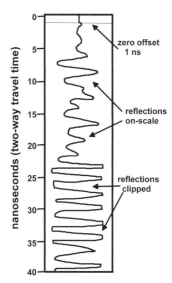

FIGURE 4.2
Trace Clipping. If range gains are misapplied, recorded waves can be "clipped," as shown in this waveform between about 25 and 40 nanoseconds.

highly reflective brick were overwhelmed with very high reflections, but these could be easily discounted as areas having intact burial features. The higher amplitudes recorded in the other areas of lower reflectivity where the graves of interest were located were then gained high enough to be visible.

If reflection data are highly attenuated with depth in the ground, often no reflections will be received from below a certain depth in the ground at all. If the time window were open to that depth, but no wave amplitudes from within the ground were being recorded, the automatic gain settings on most GPR systems will still increase whatever energy is available. Since there are no waves coming from within the ground, the range gain settings will increase only external noise that is being recorded in that portion of the time window. Anomalous and unusable data would then be recorded from those depths and must be ignored during later data analysis. Noise that is increased in this way (usually in the later portion of the time window) is generated from system noise within the GPR unit and other random interference, such as from FM radio transmission. Amplitude variations generated from differences in radar energy coupling due to changes in surface materials can also be accentuated in these later portions of the time window if no reflections from those depths are being recorded.

Once gains are set during the initial calibration of the GPR system, they should remain constant for the whole grid being surveyed with a particular antenna. If they are adjusted for any reason, the processed reflection data will display very different reflection amplitudes from the same depths in different parts of the grid, which may be confused with geologic or archaeological changes of importance. If the gain settings are changed either on purpose or by mistake and noted, it is still be possible to normalize the amplitudes of recorded reflections to achieve consistency using postacquisition data processing. Adjustment of any or all of the other setup procedures including the time window, stacking, filtering, and sampling rate will always necessitate regaining the amplitudes, as the waveform is also modified with these changes.

Vertical Filters

Vertical filters remove high- and low-frequency noise from recorded reflection traces that may be generated from system noise or frequency interference. Just as with range gaining, a school of thought believes all reflection data should be recorded in the field as raw data, and frequency filtering should only be applied during postacquisition processing. In this way all reflections, whether good or bad, are acquired. The thought is that if unfiltered data are acquired, what might be considered "bad data" can possibly be filtered and improved upon later.

The other idea is that filtering is necessary during collection so that reflections can be more easily visible and begin to be interpreted while they are being collected and are visible on the computer screen. Also, if filters are applied prior to collecting any data in the field, the optimum data quality can usually be arrived at, which might only be estimated at a later time. Some experimentation with filtering while still in the field is usually the most advantageous method, as the best data quality possible can often be collected immediately. If further filtering is necessary later on, good data can only be improved upon.

Some GPR units allow the recording of reflection data on two channels simultaneously, and therefore both raw and filtered reflection data can be acquired for each transect (Fenner 1992). Both data sets could then be processed once back in the office and compared, prior to choosing one or the other for final interpretation.

Vertical filters, also called *band-pass filters*, are employed to remove anomalously high- and low-frequency noise during data recording (Bucker et al. 1996; Fisher et al. 1994; Urliksen 1992). Terms for this filtering are *high-pass* and *low-pass*, which were originally coined for radio transmissions in the early twentieth century. The high-pass filter removes low-frequency waves (it allows the high frequencies to "pass by" a low-frequency cutoff where they can be recorded), which are often generated from "system noise" inherent to each particular radar device. These data can sometimes be seen on an oscilloscope or computer display of the recorded traces as long wavelengths superimposed on a standard reflection trace. The amount of low-frequency noise recorded will change with the antenna used, the cable length, and the type of control unit. It is usually a function of GPR system design.

Anomalously high-frequency data can be filtered out with low-pass filters (the frequencies lower than a cutoff frequency are allowed to "pass by" and are then recorded). These anomalous frequencies are usually received from FM radio transmissions or other electromagnetic disturbances nearby. High-frequency noise of this sort is easily visible when the antenna is not being moved and the generated waveform, visible on a computer display or oscilloscope, can be seen to be "flickering" because of the high-frequency noise. Because most antennas are capable of recording frequencies within one octave or so of their center frequency, a 400-megahertz antenna could be receiving energy from between 200 and 800 megahertz, or even higher in the frequency bandwidth. If a "clean" waveform is generated with high- and low-pass filters placed at 200 and 800 megahertz, then there is a good chance that high-quality reflections from within the ground will be recorded within the bandwidth of the 400-megahertz antenna and not external noise. If a good deal of noise is still being received at the antenna with these band-pass filters, it might be a good idea to decrease the low-pass filters to 600 megahertz or even lower, which would remove much of the higher frequencies, which are likely produced from nearby radio transmissions.

Care must be taken not to remove what may be reflections from within the ground during this type of filtering, but often data sets are so noisy that no coherent reflections are visible at all. One survey of note was collected in an incredibly noisy area, and it did not appear that any reflections were being recorded from within the ground at all but only noise (figure 4.3). Reflection profiles were so noisy that it appeared the survey would be a total failure. This noise was likely caused by a large amount of FM radio and other communication band noise in the urban area where the survey was conducted, which was bounded by two busy highways, near a cluster of radio antennas and very close to an airport. When both the high- and low-pass filters were narrowed to 50 megahertz on either side of the center frequency of the antenna, the background noise was effectively removed, and reflections in the ground from buried pit house floors became immediately visible (figure 4.3).

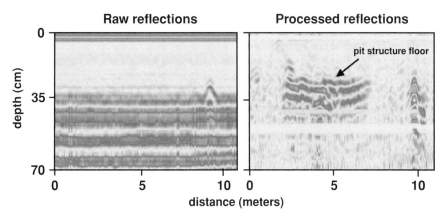

FIGURE 4.3
Frequency Filtering. Sometimes reflection data are collected that are so noisy that few usable reflections are immediately visible. When the data are frequency filtered and the background removed, important reflections from buried features are sometimes visible. These data were collected with a 500-MHz antenna in Tucson, Arizona. Almost 90 percent of the data were removed by vertical filtering, and the resulting reflections were regained to make the pit structure floor visible.

It is important to note that when vertical and horizontal filters are applied in the field prior to data acquisition, other adjustments such as the time window, sampling rate, transmit rate, and range gains must also be adjusted and possibly reset a number of times before any data are collected. All the manual adjustments noted here are part of an iterative process, and a number of experimental profiles should be collected, each with different settings applied, prior to gathering the final reflection data in transects. If good reflection data are being acquired at the necessary depths once the adjustments are set, then the settings should remain the same for all reflection profiles acquired within a grid. Often these calibration steps can take an hour or more, and it is prudent to always plan time for experimenting with all the preacquisition settings to assure the highest-quality data. The computer software included with some digital units can make many of these adjustments automatically or by employing standard stored settings set by the GPR system manufacturer. This is never a good idea. It is rather advisable to manually adjust settings for local conditions and never be lured into using preset parameters that might have worked well in the factory shop but are totally inappropriate for the local conditions encountered.

5
Velocity Analysis

One of the primary purposes of modern GPR surveys is to accurately map stratigraphy and buried archaeological features in three dimensions. In the past, many GPR studies, especially those conducted in archaeological investigations, had the limited objective of defining buried anomalies that could potentially represent archaeological features, which could be excavated later. The actual depth and orientation of any features discovered, and the nature of surrounding stratigraphy that may have been related to those features, were usually of secondary interest.

In contrast to most of these early "anomaly hunting" GPR studies, most recent work has been conducted for the purpose of noninvasively mapping buried features in detail, sometimes without ever having to excavate. Many times when excavations are planned as a follow-up to geophysical mapping, GPR maps can very accurately delineate specific areas (that are hopefully the areas of importance related to certain research questions) to concentrate on without the need for extensive digging. To accomplish these goals, precise subsurface mapping in true depth is necessary.

Archaeologists with a geological background have also learned that GPR data yield excellent stratigraphic information about sediments and soils that surround archaeological features of interest (Conyers 1995; Conyers et al. 2002; Imai et al. 1987), which are usually areas of sites that are rarely studied in detail by standard archaeological methods. This kind of stratigraphic information, unobtainable in any other way except with long trenches or dense coring, can be of great value when reconstructing buried topography, studying anthropogenic disturbance, analyzing postdepositional processes, or mapping buried soil zones or other ancient landscape features.

The changing focus of GPR exploration in archaeology has necessitated accurate subsurface mapping in real depth. Specific reflections visible in GPR data, which are always measured in two-way travel time, must usually be tied to known stratigraphy or archaeological features at measurable depths. This conversion of two-way travel time to depth can be conducted as part of the preacquisition equipment calibration procedure and at the very least must be done before any realistic data interpretation can begin once returning from the field. Conversion of radar travel times to depth can only be conducted if the velocity of the material through which the radar energy is traveling can be calculated. This chapter describes a number of field and laboratory tests that can be performed to arrive at those velocity measurements.

Radar wave travel time is the only direct measurement obtained using GPR equipment in the field. Depth (or distance) to buried interfaces or features of interest can be directly measured only using a measuring tape or other distance measuring device in an open excavation or outcrop or by probing or coring. If both time and distance are known from these two direct measurements, average velocity of wave propagation in the ground can be calculated. There are two general field techniques for determining velocity: the *reflected* (or possibly *refracted*) wave method and the *direct-wave* method. Reflected wave methods require that radar energy be reflected from objects or stratigraphic interfaces at depths that can be directly measured (Conyers and Lucius 1996; Sternberg and McGill 1995). Direct-wave methods transmit radar waves directly through the ground, from one antenna to another, also along a measured distance. Other methods of obtaining velocity are a direct measurement of the relative dielectric permittivity of samples in the laboratory (which rarely mimic field conditions) or an analysis of the geometry of reflection hyperbolas generated from buried point sources using various computer programs.

If possible, multiple velocity tests should be conducted at different locations of a study area because it is common for the velocity of soils and sediments in a test area to change both laterally and with depth. Lateral velocity variations are most commonly caused by changes in water saturation and lithology across a site. Water content is usually the single most significant variable that affects radar wave velocity (Conyers 2004). Dry quartz sand has a RDP of about 4 (table 3.1), which calculates to a radar wave velocity of 14.99 centimeters per nanosecond (equation 3.1). In contrast, the RDP of water is about 80, which yields a radar velocity of 3.35 centimeters per nanosecond. Therefore, if only a small amount of water is contained in the pore spaces of dry sand, the velocity of radar energy traveling in it will decrease significantly because of that additional moisture. In most settings, the water content of soil and sediment will naturally increase with depth, and therefore the average radar wave velocity of the material will correspondingly decrease.

The degree of residual water content in sediment and soils located above the water table, as well as the depth to the water table, can often fluctuate dramatically across an area due to changes in surface topography, stratigraphy, and the location of drainage features. In archaeological contexts, buried anthropogenic features can also create layers of different composition that affect water saturation and create dramatic velocity changes across a site. Velocity is therefore influenced by water saturation differences as they are controlled by changes in the composition of sediment and soils. Many times it is difficult to determine the causes of velocity differences across an area because they can be related to both water saturation changes and material differences, or, usually, both.

It is important to recognize that velocity measurements at a site are often valid only for GPR data that are collected within a few days (or sometimes a few hours) of when the tests are performed. Changes in velocity can vary dramatically with time as sediment and soil moisture fluctuate seasonally and can sometimes change rapidly, even during the time a survey is being carried out, due to torrential rainfall, snowmelt, or flooding. For example, velocity tests performed at a site in Central America, consisting of mostly volcanic ash, during the rainy season yielded a RDP of 12 (average velocity of 8.7 centimeters per nanosecond) (Doolittle and Miller 1992), while similar tests performed in the same area at the end of a 6-month dry season measured a RDP of about 5, or an average velocity of 13.4 centimeters per nanosecond (Conyers 1995; Conyers and Lucius 1996). In this case, if the velocity tests performed during one season were used to process and interpret GPR data acquired just a few months later, the velocity adjusted depths of radar reflections would be extremely inaccurate. The same kind of dramatic changes have also been seen overnight. In the American Southwest, good radar reflections were obtained one day, from depths approaching 2 meters with a 500-megahertz antenna. Overnight 3 inches of rain fell, and the next day, poor reflections from a maximum depth of only 50 centimeters were recorded, with a completely different calculated RDP (Conyers and Cameron 1998). In this case, the addition of water changed not only the velocity of radar propagation but also the depth of radar energy penetration in the ground.

REFLECTED WAVE METHODS

The most accurate and straightforward method to measure velocity is to identify reflections in GPR profiles that are produced from objects, artifacts, or zones of interest, which occur at known depths. These methods allow for a direct determination of the average velocity of radar waves from the surface antenna to a measured depth. In the past, these types of velocity tests have been conducted at archaeological sites on objects as diverse as buried whale bones (Vaughan 1986), copper wire (Kenyon 1977), and empty paint cans (Doolittle and Miller 1992). Because metal is a near-perfect radar energy reflector, the reflections generated in a profile crossing a metal object are easily identifiable on most GPR profiles as distinct hyperbolas. Other tests of a similar sort can be performed when a buried wall or some other point source reflection feature is partially exposed in an excavation and can be identified in GPR profiles (Conyers and Lucius 1996). In a similar way, identifying a distinctive reflection generated from a noticeable material change in the ground, and then coring or excavating a test trench to expose it, will serve the same purpose as long as the exact reflection surface can be positively identified. This was done where a stream channel, with a distinctive base, was seen in reflection profiles and this surface could be easily distinguished in an auger probe as a change from coarse sand and gravel to clay at the base of the channel (figure 5.1). The depth of the channel bottom was then measured in the auger hole and velocity could be easily calculated.

In all cases, velocities measured in this way are an average from the ground surface to the depth of the object or interface measured. Multiple tests of this type from many depths in the ground can determine if there is a large change in

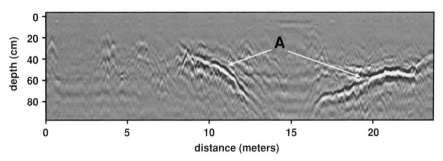

FIGURE 5.1
Stratigraphy Identification for Velocity Determination. The base of a sandy river channel (A) was visible in profiles collected near Lompoc in coastal California with a 500-MHz antenna. The depth to the base of the channel was tested with an auger, and velocity was calculated.

velocity with depth and also laterally across a site. Without accurate velocity measurements, all interpretations of GPR profiles that require any three-dimensional resolution will be speculative.

The easiest and most accurate method for determining velocity is to excavate or find a nearby trench or outcrop and place an iron bar horizontally in a vertical face. Antennas are then slowly pulled over the bar while subsurface reflections are recorded in a profile. The metal bar will be apparent as a distinct reflection hyperbola, such as that in figure 3.15. To obtain the maximum amount of reflection from a thin metal bar, the long axis of the surface antennas (from side to side) must be oriented parallel to the length of the horizontal bar. This antenna orientation will create an electric field that is also oriented parallel to the bar, producing the maximum amount of reflection. Also, when doing these types of tests the correct antenna frequency suitable for the depth and object size that needs to be illuminated must be used. If a low-frequency antenna is used, a fairly large target may be necessary in order for it to be visible on standard two-dimensional profiles, because of this antenna's lesser resolution.

When reflection profiles are immediately visible on the computer screen during data collection, and if a time scale is superimposed on the profile, the apex of the hyperbola can be measured in time. Time and depth will then yield velocity, which can be immediately used to calculate the time window for the depth necessary to resolve features of interest. If the iron bar is not visible, the profile may have to be processed later on to increase the visibility of the target, using the data-filtering and enhancement methods discussed in chapter 6.

When multiple tests are made at different depths at a study area, differing velocities will often be calculated because of changes in water saturation and ground composition with depth. If only one average velocity is used (which is often the case), there could be vertical distortion if a velocity derived from a shallow direct-wave test were used (which would likely have a relatively high velocity) to depth correct all reflections in a profile. Reflections produced from interfaces close to the ground surface would be corrected to about their correct depth, but those from deeper in the ground would likely appear too shallow, as radar travel times from deeper in the ground are probably slower. The alternative would also be true if an average velocity obtained from an object deeper in the ground were used, which would yield a slower average velocity. That slower velocity, when applied to a whole data set, would tend to make shallow reflections appear deeper than they really are.

This is one of the pitfalls encountered using an average RDP (velocity) to correct radar travel times to depth during data processing. If different RDP values were known for specific depths, it is possible for some computer programs to convert profiles measured in time to depth profiles using varying RDPs for different depths. There might still be some distortion in the resulting profiles at the boundaries between units with differing RDP because abrupt velocity changes would be inferred that may not be real. A more accurate approach would be to obtain RDP values at varying depths and then compile a velocity variation curve for the stratigraphic section as a whole. Some GPR data-processing programs allow this type of sophisticated correction to create more accurate depth profiles.

It is quite possible that the imposition of only one velocity (or RDP), derived from a test performed in only one area of the site, to all of the GPR data acquired in surrounding grids could yield spurious depth calculations if subsurface conditions change laterally. For instance, if there were higher water saturations in one area because of a perched water table (yielding slower velocities), actual depth to interfaces of interest might also vary considerably. For the most part, without detailed stratigraphic knowledge of a site, these variations would probably go undetected, and interpretations regarding the depth to certain units would be inaccurate. A more geographically dispersed data set of velocity calibration tests might possibly allow a velocity gradient map to be made in three dimensions and produce more accurate profiles and maps. Without them, these potential velocity problems must be accepted as part of the inherent imprecision of the GPR method.

Another type of mistake that is possible with velocity conversions was made in an area where the only place available for a direct-wave test was an excavation face that had been left exposed to the elements for two years (Conyers et al. 2002). This occurred in a desert area in Jordan where wind blown sand was the matrix material of the site. Numerous direct-wave tests were made on an iron bar at different depths along that exposed face and velocities were found to be consistently very high throughout the section, with an RDP of about 3. All reflections in the grid were then corrected using this average velocity, and the depth to features and stratigraphic interfaces of interest throughout the study area were mapped. When these features were later excavated, it was found that all mapped features were almost three times shallower than predicted in the GPR maps! After evaluating what could have gone wrong, it was concluded that the sand in the exposure where the velocity tests were performed had been allowed to dry out significantly along the excavation face, creating an anomalously dry material that allowed radar waves to travel at a very high velocity. The same type of material that remained buried (where the GPR data were collected) retained its natural moisture. A very erroneous velocity was therefore applied to all the reflection data. This simple, but potentially disastrous, mistake immediately called into question the capability of the geophysical archaeologists conducting the survey, as the features were uncovered while they were still present in the field (Conyers et al. 2002). Fortunately, all involved understood (or were made to understand!) the potential velocity pitfalls in GPR processing and were able to quickly correct the mistakes in processing before additional excavations were conducted. If excavations had taken place days or weeks later, after the geophysicists conducting the survey had long left the field, all the GPR survey results might have been called into question. In this case velocity tests should have been conducted in newly excavated trenches, not the older excavations where the sediment had been allowed to dry out.

DIRECT-WAVE METHODS

Although sometimes not as accurate as reflected wave methods, direct-wave techniques provide an additional way to determine radar wave velocity in the field. In these types of tests, two antennas are separated, with the material to be tested located between the two. One antenna then transmits to the other, and the one-way transmit time between the two can be measured. If the distance between the two antennas is known, velocity can be calculated (Conyers and Lucius 1996). One type of a test of this sort is called *common midpoint* (CMP) (Fisher et al. 1994; Leckebusch 2003; Malagodi et al. 1994; Tillard and Dubois 1995), and a similar test is *wide-angle refraction and reflection* (WARR) (Imai et al. 1987; Milligan and Atkin 1993; Reynolds 1997: 710). *Transillumination* tests are a third type, all of which are based on the same general method where radar waves are transmitted in a one-way direction between two antennas that are separated by the material to be tested.

In all of these types of tests, a GPR system that allows two antennas to be separated is necessary. Many GPR systems commonly used by archaeologists collect data with dual antennas that are permanently attached, recording on only one channel with one antenna cable leading to the control unit, and for this reason these types of velocity tests are not as common as others. Some manufacturers produce dual antennas that are clipped together and can be easily separated, but special connections are usually necessary to allow this separation (or two cables: one connecting each antenna). Otherwise a cable splitter with two separate antennas, or a multichannel GPR system are necessary. If the only two antennas available for a direct-wave test are not the same frequency but are close (e.g., 400- and 500-megahertz antennas), both may be used, as they will both transmit and receive within each other's bandwidth.

CMP and WARR Tests

In both WARR and CMP tests, radar energy is sent from one antenna to the other as they are moved an increasing distance apart. Individual reflection traces are usually collected (and often stacked to improve their quality) in step mode for this type of test. The radar waves moving between the antennas will pass through both the air and near-surface layers of the ground and be received at the other (figure 5.2). If the distance of separation is known and the radar wave travel paths can be deduced, the arriving waves can be measured in time, and a series of velocity measurements of different layers in the ground can potentially be calculated. In the CMP method, both antennas are first placed next to each other on the ground, and one reflection trace is collected. There may be three or more wave arrivals collected at this location: one that travels in air, one along the air–ground interface and possibly more that are reflected and refracted from buried interfaces in the ground. The antennas are then separated a measured distance (10 centimeters, perhaps), and another reflection trace is recorded (figure 5.3). This collection procedure is repeated many times until the antennas are separated by as much as 5 or 10 meters. Energy will continue to travel a number of paths between the two antennas, and if the wave arrivals that have traveled within the ground can be identified, and the distance between the two antennas is known, velocity can be calculated. The same type of recording can also be made in continuous data acquisition with the imposition of fiducial marks to yield distance as measured along a tape measure on the ground.

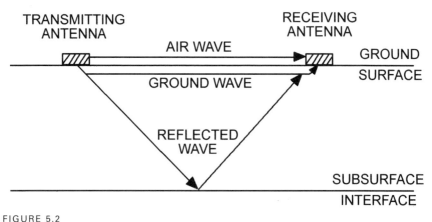

FIGURE 5.2
Travel Paths for a CMP Test. At least three waves are recorded in CMP and WARR tests: the air wave that travels in air from transmitting to receiving antennas, the ground wave that moves along the interface between the ground and the air, and one or more reflected (and sometimes refracted) waves that move within the ground.

VELOCITY ANALYSIS   107

FIGURE 5.3
Conducting a CMP Test. Antennas are separated and the transmitter and receiver are moved apart in steps (pin flags denote the antenna offset distances). Reflection traces are recorded at each location.

Reflection traces are collected in the same way in the WARR method, except only one antenna is moved while the other remains stable. Sometimes WARR tests must be performed instead of CMP tests if there are not enough people in the field at the time of the test to move both the antennas simultaneously.

In WARR tests, there should be subsurface layers to reflect energy that are either horizontal or not dipping greatly (Reynolds 1997: 711). This is because energy must move in a predictable way between the two antennas, which is rarely the case. For this reason, CMP tests are the preferable method, as the midpoint between the antennas remains the same, allowing the points where energy are reflected on each subsurface reflector to be determined at each offset location and thus aerial consistency at depth is not a requirement. Also, for the result of WARR tests to be accurate, it must be assumed that the velocity of the individual layers does not change dramatically laterally, which is also rarely the case.

Common midpoint and WARR data are typically displayed in a standard GPR reflection profile, with the antenna separation distance on the horizontal axis and time on the vertical axis (figure 5.4). As the antennas are moved apart, the first wave received is the air wave. Ideally it is recorded at time zero when the distance of antenna separation is zero, or the zero offset can be later taken into account during data processing. The second wave arrival is usually the ground wave that travels along the ground–air interface and is recorded soon after the air wave. The third, and any subsequent arrivals, are usually reflected or refracted waves generated at subsurface interfaces. In areas with shallow water tables, it may be possible to distinguish between what are referred to as "dry" ground waves and those that are "wet" and that may be traveling within saturated ground near the surface (Fisher et al. 1994). The ground wave should also intersect the air wave at time zero on the portion of the radar profile where the two antennas were touching (figure 5.4). Other radar waves that traveled deeper within the ground can be refracted within soil and stratigraphic units and sometimes reflected between subsurface layers before arriving at the receiving antenna, creating what can potentially be a confusing series of recorded wave arrivals. These subsequent arrivals can usually be differentiated from ground or air waves because they do not intersect at time zero when the two antennas were touching.

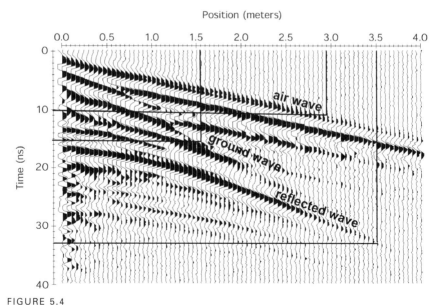

FIGURE 5.4
A CMP Test. This series of reflection traces when stacked and then viewed in a profile show the air wave, ground wave, and reflected wave from within the ground, arriving at different one-way travel times. Using these measurements both time and depth can be determined in order to arrive at velocity.

When time is measured and distance (between the antennas) is known, and the reflections from certain distances in the ground can be identified, velocity can be calculated. In figure 5.4, the air wave can be identified at any point along the wave diagram by taking a time and distance reading and calculating velocity. In this example, the air wave arrival at an antenna separation of about 3 meters occurred at 10 nanoseconds, which is a velocity of 0.3 meters per nanosecond (exactly the speed of light in air). The ground wave arrived at 15 nanoseconds when the antennas were separated 1.6 meters, which is a velocity of about 0.1 meters per nanosecond (a decrease from the air wave velocity of about a third). The reflected wave from some unknown horizon in the ground arrived at 33 nanoseconds when the antennas were separated 3.5 meters, also giving a velocity of about 0.1 meters per nanosecond.

In most cases, CMP and WARR tests usually only measure velocities of surface soils or other material that is located very near the ground surface. They should not be viewed as a way of determining velocity at any great depth unless it is somehow possible to identify actual paths of radar waves that traveled to deeper within the ground. For deeper velocity determinations, direct-wave methods should be used. Many GPR processing programs have software routines available to quickly process data collected in this way and average velocity at a number of depths in the ground can be estimated for later data processing.

A modified CMP data collection method was used as a way to collect a very precise three-dimensional reflection data set in a small area where accurate volumes of reflection data in real depth were needed (Pipan et al. 1996; Pipan et al. 1999). Using the CMP method within an 8-by-8–meter grid, many hundreds of CMP tests were performed, collecting thousands of traces that were then stacked and processed into a three-dimensional cube. This innovative approach, although difficult to conduct and time-consuming to process, produced superior three-dimensional images of the ground not available in any other way.

Transillumination Tests

The transillumination method is another direct-wave velocity test that is applicable to archaeological settings because it can be conducted in two nearby excavations where the section of material to be tested is preserved between them. It was originally developed as a method for evaluating the integrity of intact structures for both engineering and archaeological applications (Bernabini et al. 1994). To conduct transillumination tests, the faces of excavations should be as close to parallel as possible. It is also best if tests are performed soon after the material is exposed in the excavations so that any evaporation or seepage of water along the faces does not significantly change the water saturation characteristics of the material to be tested.

Two antennas, one to send and the other to receive, are then placed on the walls of the two excavations, pointing toward one another (figure 5.5). It is important that the two excavations be separated by at least one wavelength or so of the center frequency of the antenna being used to transmit so the receiving antenna is beyond the near-field zone of the transmitting antenna.

A series of transillumination tests are then made starting at the base of the excavations and moving upward. The two antennas can be moved upward either in steps, collecting one or many stacked reflection traces at each step, or continuously as radar energy is transmitted between the two. Care must be taken to keep the antennas separated a known distance and a known height from the base of the excavation as they are moved. If the antennas are moved in steps it is important that each antenna be moved the same distance from the top or bottom of the exposed faces so that the distance between the two is always known. If the walls of the excavation are sloping, then a series of distance measurements must be made in order to arrive at the antenna separations for each of the steps where data are collected.

When the material to be tested in this way is highly stratified, it is important that the electric field generated by the dipole antenna be oriented parallel to the bedding planes. To do this, the long axes (from side to side) of the antennas must be placed parallel to the bedding planes. In this way the majority of the electrical portion of the electromagnetic field will vibrate parallel to the bedding layers and there will be maximum isolation of the radar beam within each stratigraphic unit (Conyers and Lucius 1996). The cone of illumination of the radar antenna may, however, still transmit radar energy into adjacent layers regardless of the orientation of the antennas, and, as always, energy travel paths in the ground are difficult to predict or define after the fact. In all cases, the transmitting and receiving antennas must be oriented in the same direction so that there is maximum "communication" between the two.

In one transillumination test of this sort, two excavations exposed a thick section of volcanic ash (Conyers and Lucius 1996). Eight different ash units were identified on the faces of the excavations, and reflection traces were collected at each bed in seven steps, from the base of the excavation upward. A 300-megahertz antenna was put on one side of the exposure to receive and a 500-megahertz antenna on the other to transmit. An analysis of the travel times at each step showed that velocity increased from about 6 centimeters per nanosecond at the base (RDP of 23.2) to 28 centimeters per nanosecond (an RDP of 1.1) at the top (figure 5.6). The two velocity measurements at the top were probably

FIGURE 5.5
Conducting a Transillumination Test. When conducting this test, two antennas were held on parallel vertical faces, pointing toward each other and separated by the material to be tested.

not measuring radar wave travel in the ground, as radar energy probably "leaked" over the top of the excavation and traveled in the air between the two antennas, producing the RDP of 1.1, which is almost that of the velocity of radar transmission in air.

The identification of the "leaked" air waves that traveled between the two excavations illustrates the importance of using fairly deep excavations when conducting transillumination tests. If tests are conducted too close to the surface, radar waves will travel over the top or around the material to be tested, and the first arrivals may be only air waves. It is also critical that accurate distances between antennas at all positions be obtained so that air wave calculations can be made and their arrival times calculated in advance and then identified.

Knowing the horizontal separation of the antennas and the one-way travel time of the radar energy between the two antennas at each step, velocities can be calculated and RDP determined, using equation 3.1. When the velocity measurements at each of the seven steps as shown in figure 5.6 were plotted against the depth of the antennas, a velocity gradient graph was constructed. Velocity information derived from transillumination tests can be of great importance because it identifies velocity changes as a function of depth, which is not usually possible in direct-wave methods. In the graph in figure 5.6, the velocity increases at a fairly constant rate with greater depth, probably indicating gradually increasing residual water saturation, which was also visible as minor color changes in the exposed section of volcanic ash tested. The minor change in the velocity gradient at 100 centimeters may indicate a change in velocity between layers of very different composition and therefore differing water saturations.

If changes in velocity can be correlated with lithology or other compositional or water saturation changes in the material, they can yield important information for interpreting nearby reflection profiles, as it is always important to understand the origins of reflections when attempting to understand GPR data from any site. Because all reflections are generated at buried interfaces where there are distinct velocity changes, transillumination tests can be one of the best methods with which to understand these variations and possibly correlate reflections to known stratigraphic units in the subsurface.

Data from transillumination tests should still be used with caution because radar wave travel paths within the material being tested can never be known for sure, just as with CMP and WARR tests. In all these types of direct-wave tests, radar energy will tend to travel preferentially within the highest-velocity material, and the time of the first arrival that is being used to calculate the velocity may be from the waves that traveled in the "fastest" material, not necessarily those within the material from the depth at which the antennas are placed. Any wave arrivals that may have traveled through the lower-velocity layer would then be overwhelmed, obscured, or otherwise unrecognizable in the resulting traces.

Transillumination, CMP, and WARR tests should always be performed in conjunction with direct-wave tests of objects at known depths. The combination of both types of velocity test methods will yield both average vertical velocity measurements as well as a velocity gradient with depth.

LABORATORY MEASUREMENTS OF RDP

At most archaeological sites, samples of subsurface units can usually be collected for later processing in the laboratory to determine relative dielectric permittivity, electrical conductivity, and magnetic permeability. These measurements can

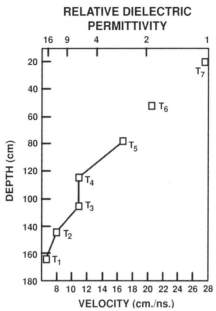

FIGURE 5.6
Transillumination Test Results. Velocity measurements can be determined as a function of depth. In this test there is a gradual decrease in velocity with depth, because of higher water saturation in the material being tested. Tests 6 and 7 measured energy that traveled in air over the top of the excavation.

then be used to estimate velocity and energy attenuation for materials at a site. They are also valuable when constructing two-dimensional computer models, discussed in chapter 7. If soil and sediment samples are collected and immediately stored in watertight containers, they can be considered as approximating field conditions. In reality, however, samples that are stored and transported in plastic bags or bottles will never replicate field conditions as their porosity, grain packing, and water saturation will all change somewhat in the process of gathering and transporting them.

Unfortunately, only a few devices can make these types of laboratory measurements, none of which are readily available to most archaeologists. One way of determining the magnetic and electrical properties in the lab is by using techniques described by Olhoeft and Capron (1993) and Saarenketo (1998). In these tests, samples are first dried and crushed and then subjected to differing frequencies of electromagnetic energy in a device called a network analyzer (Conyers 2004). Measurements of RDP, conductivity, and magnetic permeability can be made when the samples are totally dry and at differing water saturations. Water saturation changes that might be found in the field can be simulated by progressively wetting the samples with water from a dropper between tests, allowing time for the water to adequately penetrate the material.

There is a danger when using data from these types of laboratory tests because changes in the material that affect grain packing and porosity always occur during the testing procedure (Olhoeft 1986). Also, when water is artificially placed back into a sample after it has been artificially desiccated in the laboratory, conditions unlike those in the field are created. Devices that measure the electromagnetic properties of samples also tend to transmit more energy into a sample than would usually occur in the field, where attenuation and wave dispersal with depth always occur.

A laboratory test of this sort was conducted on what appeared to be a clay soil from central Illinois (Conyers 2004). It was later determined to be a silty clay, and the clay was determined to be not mineralogic clay but wind-blown clay-size rock fragments. The sample was first totally desiccated in an oven overnight to remove all residual water. Its relative dielectric permittivity was then measured at many frequencies ranging from 10 to 1,200 megahertz (figure 5.7). An analysis of those readings indicated that these types of measurements are somewhat frequency-dependent with variations at the high and low end of the range, but RDP was fairly stable within the frequency range of most GPR antennas (200 to 1,000 megahertz). These types of laboratory measurements are always frequency-dependent. At high frequencies, some electromagnetic energy is lost in the atomic structure of the materials due to displacement currents, caused by small perturbations within the orbits of electrons. At low frequencies, ions in the material cannot respond fast enough to the imposed electromagnetic field, and there is greater ionic conductivity, and therefore higher RDP values are measured.

Those frequency variations aside, what is most remarkable about the test shown in figure 5.7 is how sensitive the samples were to the addition of water. When only 0.5 cubic centimeters of distilled water was added to the total sediment sample of 13 cubic centimeters, the RDP increased from 3 to about 8. And with an additional 0.5 cubic centimeters of water it increased even more, to about 17. This test dramatically illustrates how important the addition of a small amount of water is to the velocity of radar travel (and therefore RDP of the material in the ground). This same phenomena is also visible in the transillumination test in figure 5.6 where ash beds of identical composition have greatly varying RDP values with greater depth in the ground (and therefore greater wa-

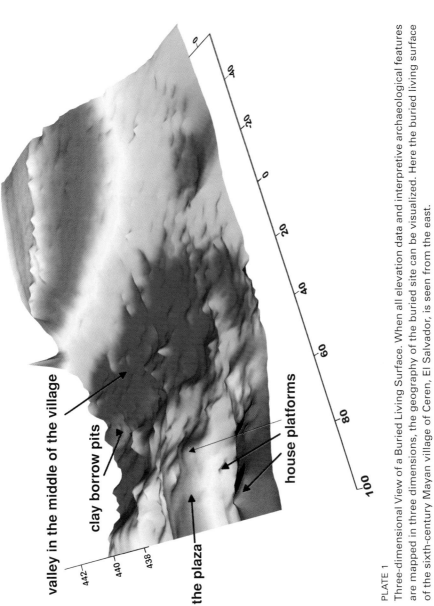

PLATE 1
Three-dimensional View of a Buried Living Surface. When all elevation data and interpretive archaeological features are mapped in three dimensions, the geography of the buried site can be visualized. Here the buried living surface of the sixth-century Mayan village of Ceren, El Salvador, is seen from the east.

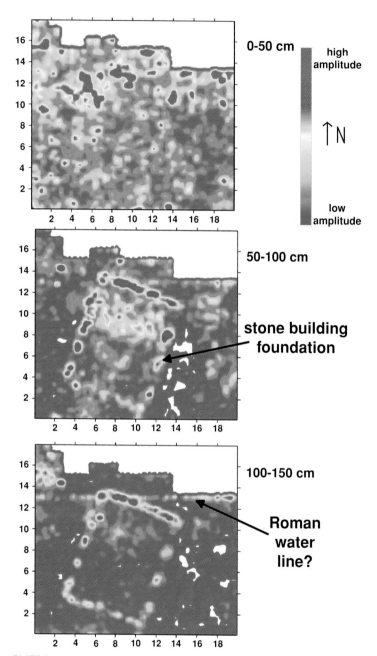

PLATE 2

Amplitude Slice Maps. These amplitude slice maps illustrate a portion of the Lower Market garden at Petra, Jordan. Each slice is about 50 centimeters thick. In the upper slice only near-surface stones are visible, but in deeper slices a distinct stone foundation is visible, with a possible Roman water line below it.

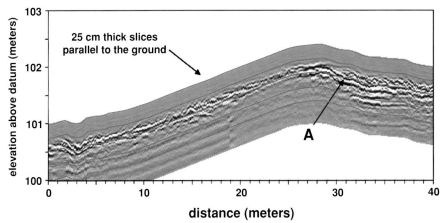

PLATE 3
Slices Crossing Subsurface Bedding Planes. Slices are most often constructed parallel to the ground surface, in this case every 25 centimeters in the ground. When those slices cross bedding planes, as at location A, an amplitude change will be registered producing a false anomaly in that slice, as seen in the map in plate 4. This profile is number 9, whose location is shown in plate 4.

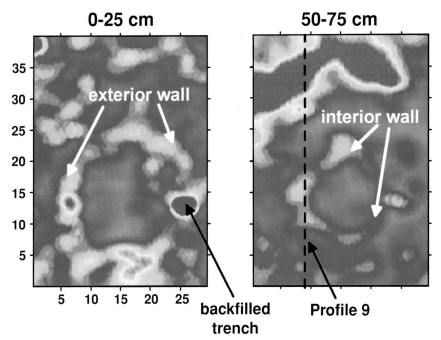

PLATE 4
Anomalies Created by Slices Crossing Bedding Planes. The large red amplitude anomaly in the 75- to 100-centimeter slice is a product of the slice crossing the buried stratigraphic layer that is seen in profile 9 in plate 3. The other interior and exterior walls of this kiva in southeastern Utah are visible as high amplitudes.

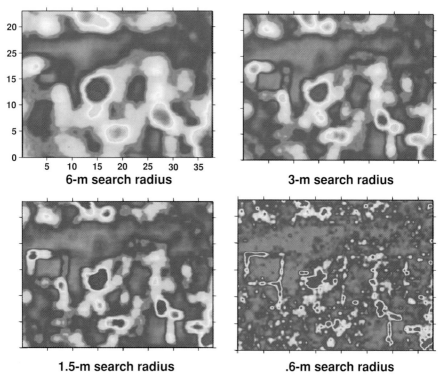

PLATE 5
Interpolation Differences in Slicing. Amplitude slice-mapping program commands allow different interpolation radii that can make the difference between being able to visualize features or not. If too large a search radius is used (6 meters), little is visible in this slice from 75- to 100-centimeter depth at the Fort Garland Site in Colorado. A 0.6-meter search radius allows the buried walls to be clearly visible.

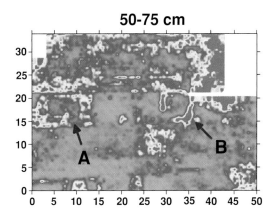

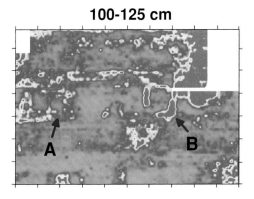

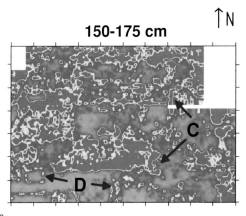

PLATE 6

Amplitude Slices on Horizontal Ground. These three amplitude slices are from differing layers in the ground, at a historic site in Albany, New York, which is today a paved parking lot. The upper two slices show the remains of nineteenth-century domestic structures (A) and a kiln for malting grain (B), both of which are known from historic maps. In the deepest slice, there are older features (C and D), which have no historic records to indicate their function.

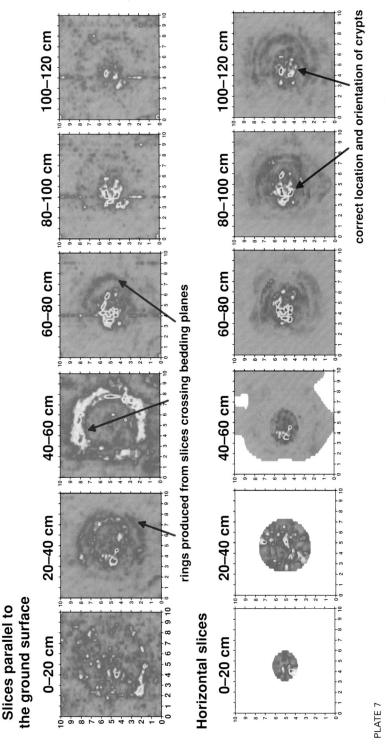

PLATE 7

Amplitude Slices on a Mound. Slices that are produced parallel to a ground surface on this simulated burial mound cross bedding planes and produce anomalous amplitude readings at those intersections. When profiles are adjusted for topography first and then the grid as a whole is sliced horizontally, more accurate spatial placement of amplitudes occurs, and the high amplitudes are showing the correct location of the burial crypts within the mound.

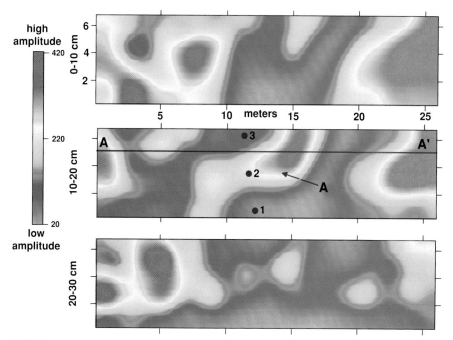

PLATE 8
Amplitude Slices Illustrating Subtle Ground Features. When many profiles (including that shown in figure 7.10 along cross section A-A') are analyzed for amplitude and mapped spatially, trends in amplitude can often yield information about the nature of buried units. In this grid, a small shallow creek is visible as a high-amplitude sinuous reflection (A), which was confirmed in auger hole 2. Auger holes 1 and 3 recovered marsh sediments.

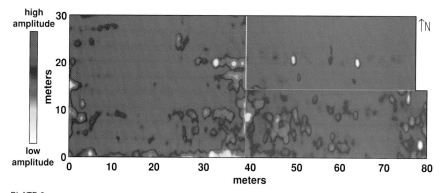

PLATE 9
Grave Amplitude Slice Map. This map is denoting areas of anomalously low reflections derived from grave shaft fill in a cemetery near Boulder, Colorado.

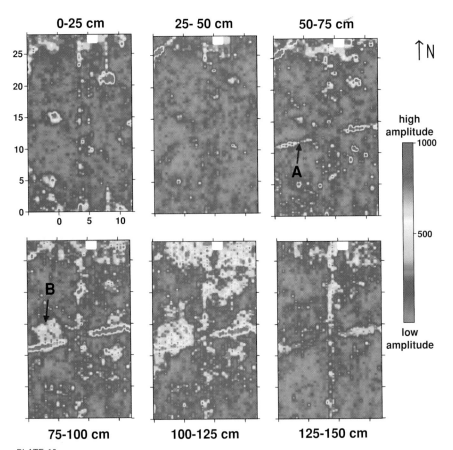

PLATE 10
Amplitude Slice Maps of a Very Subtle Floor Feature. These slices show a pipe (A) and an adjacent square outline of a moderately high-amplitude feature (B) in the 75- to 100-centimeter slice. The linear feature was known to be a plastic water pipe. The square feature was excavated and found to be a very thin sand layer that was at one time the floor of a historic building.

ter saturation), as well as with porosity changes between different stratigraphic units (that also changes the amount of water held in any one stratum).

An interesting approach to laboratory measurements was taken by Sternberg and McGill (1995) in Arizona. At their sites, samples were taken of subsurface sediment units, which were analyzed for their particle size, mineralogical constituents, and water saturation. Without having to rely on laboratory measurements of RDP and electrical conductivity, they compared their mixtures consisting of sand, clay, and water to published tables of RDP and other measurements for those materials (Olhoeft 1986) in a fairly simplistic but effective way. Relative dielectric permittivity and velocity were then estimated for their unique field conditions, and radar times were converted to approximate depth, with good results.

ANALYSIS OF POINT SOURCE REFLECTION HYPERBOLAS

When GPR transects cross subsurface features that generate point sources, such as pipes, walls, rocks, or small void spaces, hyperbolic reflections are generated (figure 3.15). The geometry of the generated hyperbola (in general how steeply the arms of the hyperbola dip) is a function of the average velocity of the material

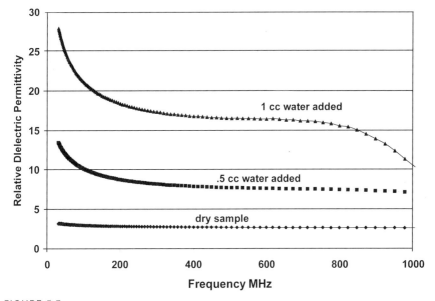

FIGURE 5.7
Laboratory Tests of RDP. The sample of clay when dry had an RDP of 3, which increased rapidly as small amounts of distilled water were added, illustrating how important water is to the transmission and reflection of radar energy in the ground.

through which the radar energy passes (Leckebusch 2000; Lucius and Powers 2002; Martinaud et al. 2004; Paniague et al. 2004; Tillard and Dubois 1995) and the size of the point source that generated the hyperbola. When a profile is collected with equally spaced surface fiducial marks (or with a survey wheel where reflection traces are equally spaced along a transect), distance is known in one dimension, and the velocities of many wave travel paths to and from the point source in the ground are measured when the antenna is at different surface locations. A computer-generated hyperbola is then "fit" to the hyperbola generated in the ground, and its dimensions are calculated (figure 5.8). Velocity and distance can then be measured, and average velocity from the ground surface to the point source can be calculated using computer programs that apply trigonometric functions to these measurements. The simplest version of this type of processing program is called Fieldview (Lucius and Powers 2002), but there are many other commercially available programs that perform the same calculations.

A GPR survey that generates many hyperbolas in reflection profiles can be readily used to evaluate velocity with both depth and throughout a grid without having to resort to any of the above field velocity test methods. In this method, hyperbolas can be tested in many reflection profiles at different depths in the ground (assuming there are hyperbolas present) and variations in velocity can be determined both with depth and spatially across a grid. Often this is more information than is needed, or even wanted, and usually an average velocity for the data set as a whole is sufficient to correct radar travel times to approximate depth.

VELOCITY ANALYSES CONCLUSIONS

The most accurate velocity tests are those performed in the field that directly measure the radar travel times of objects at known depths. The object to be resolved should be metal in order to maximize radar reflections. If possible, one or more velocity tests should be made within or near a proposed GPR grid to test the velocity of all materials that will potentially be studied. At depths greater than a few meters, iron bars or other relatively small objects may not be visible, and larger objects, such as a buried structure wall, can be used as a target. Without these types of tests, the correlation of important stratigraphic units or buried cultural materials to reflections measured in radar travel times will always be suspect unless the stratigraphy is so simple and the archaeological features are so dramatic that the origin of resulting reflections is not in doubt. This is rarely the case.

If two or more excavations are available in close proximity, transillumination tests can be performed. The velocity data gathered from these types of tests can

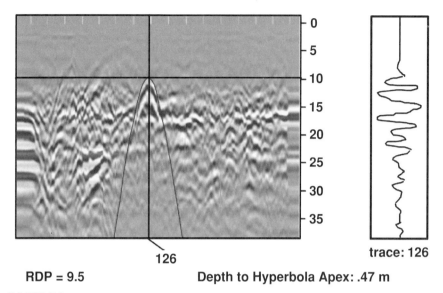

FIGURE 5.8
Hyperbola-Fitting Velocity Tests. Computer programs can be used to fit hyperbolas of a known geometry to reflection hyperbolas in radar profiles in order to arrive at velocity.

yield velocity gradient curves and also delineate interfaces in the subsurface most likely to reflect radar energy. These types of data can be especially valuable in correlating reflections to known stratigraphy and other subsurface features.

If excavations are not available at a site, velocity of near-surface zones can be estimated using common midpoint or wide-angle reflection and refraction tests. These tests can be used to estimate near-surface velocities, but are usually not as valuable in obtaining velocity information from any great depth.

It is important to understand that any data derived from field velocity tests of any sort must be applied only to GPR reflection profiles that were acquired at about the same time. Ground conditions can change with the season or due to other factors such as heavy rainfall or snowmelt and subsurface radar velocity can change accordingly.

Lacking field velocity tests of any sort, samples of overburden material collected in the field can be analyzed in the laboratory or compared to standardized tables in order to obtain electrical and magnetic properties. These data can be used to arrive at an approximate relative dielectric permittivity, from which velocity can be calculated, if no other information is available.

A GPR grid that contains an abundance of reflection hyperbolas generated from buried point sources can also yield important velocity information by using computer programs that fit the geometry of hyperbolas. These analyses can be performed without the need for excavations and are often quite accurate and an easy way to obtain average velocity estimates.

All or some of the velocity tests described in this chapter should be performed as a matter of course during GPR surveys. For the most part, they are not difficult or time-consuming and will yield valuable time–depth information that is necessary in order to process raw GPR reflection data. Reflections that are recorded and interpreted only in time can only be used as a crude estimate of depth without accurate time–depth conversions.

6

Postacquisition Data Processing

Usually GPR reflection profiles that are viewed during or directly after being acquired in the field are obscured by what interpreters refer to as "noise," "reverberations," "interference," "multiples," "clutter," "spikes," and "snow." Often reflection profiles can contain extraneous reflections such as air waves, multiple reflections, and point source reflection hyperbolas, which also make them difficult to interpret or process into usable maps. Raw field reflection profiles are also usually not collected with either accurate depth or horizontal scales, which must be placed into the reflection records after returning from the field. In almost every case, "raw" reflection data must be "cleaned up" and adjusted in some way prior to interpretation.

A large number of commercial, proprietary, and free GPR processing programs are available for this type of data processing. Many of the GPR processing techniques included in these programs have been partially borrowed and then modified from those used in the petroleum industry, which processes seismic reflection data, or other remote sensing applications that deal with complex image processing (Hatton et al. 1986; Malagodi et al. 1996; Milligan and Atkin 1993; Sheriff and Geldart 1985; Ulriksen 1992; Yilmaz 2001). In all cases, to process GPR reflections the original data must either have been recorded digitally or be analog data that have been digitized.

One should never attempt to use processing programs "off the shelf" without understanding the implications of each data manipulation technique. That is because many techniques were written for very specific objectives, and some or

Table 6.1. Common Postacquisition Processing Objectives and Methods

Processing Objective	Methods to Be Used
Correct vertical and horizontal scales in reflection profiles; average or smooth reflections along transects	Standard "rubber sheeting"; distance normalization and vertical exaggeration; trace stacking
Remove horizontal banding created by system noise and frequency interference	Background removal; vertical high-pass filters
Remove high-frequency noise ("snow")	Low-pass frequency filters; F-k filters
Remove multiple reflections	Deconvolution
Remove and compress point source hyperbolas to their sources; correct the orientation of steeply dipping layers	Migration

even all may not be applicable to one's archaeological or geological objectives. It is always dangerous to process data only in the hope that the final product will "look better," without understanding exactly what the processing step is doing to the original field recordings. A list of the common data processing techniques that can potentially be used for GPR data and their objectives is found in table 6.1.

As with most postacquisition processing, some steps should be performed in a certain order (Pedley and Hill 2003; Woodward et al. 2003). In table 6.1, these steps are listed in the approximate order that they should be attempted. For instance, reflection profiles should always be corrected spatially in the horizontal and vertical dimensions prior to initiating other steps. Background removal may then remove horizontal reflections in profiles but cause a decrease in the remaining amplitudes of some reflections, which then need to be increased with range gaining to be visible. Or background removal may eliminate many reflections leaving a good deal of "snow," which must then be removed with frequency filters. Once this step is performed, a third procedure, range gaining, may then may be necessary to enhance important reflections not otherwise visible. Any number of these types of incremental steps may be necessary, often in different orders of application.

SCALE CORRECTION AND THE CREATION OF REFLECTION PROFILES

The simplest processing procedure takes the individual reflection traces from each transect and places them in sequential order so that they may be viewed as a two-dimensional vertical profile through the ground. These can be printed in "wiggle trace" format, which shows the individual traces and their amplitudes, or in a gray scale interpolated image, where amplitudes of individual reflections vary in shades of gray (figure 6.1). In the same step, reflection profiles

POSTACQUISITION DATA PROCESSING

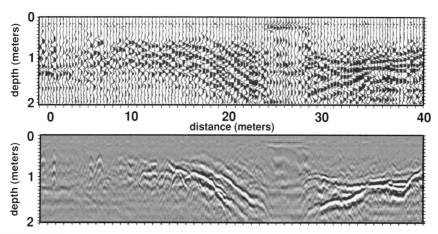

FIGURE 6.1
Wiggle Trace and Gray-Scale Reflection Profiles. Individual traces collected in a transect can be displayed in wiggle trace format (upper profile) or in a gray-scale image that tends to blend all reflection traces together for a smoother look (lower profile).

can be exaggerated in both the vertical or horizontal dimension to emphasize certain aspects of the stratigraphy or buried archaeological features. This exaggeration can often be accomplished with a mouse while profiles are visible on the computer screen, which will expand scales quickly until a desired look is achieved. A series of these fairly simple processing steps, once they have been determined, can then be applied to all reflection profiles within a grid automatically.

Standard reflection profiles can also be modified so that the relative amplitudes of reflections are assigned colors. In this way, significant reflections that may represent important interfaces in the ground are more readily visible to the human eye. Care must be taken in choosing a color palette, however, because sometimes many-colored reflection profiles can be "busy" and difficult to interpret (Leckebusch 2003; Milligan and Atkin 1993), and gray scale often is easier for the human brain to interpret than many complex colors.

If reflection data are collected in continuous mode, there will always be an unequal number of reflection traces between fiducial marks because of the inconsistencies in the antenna towing speed. Distance normalization commands will "rubber-sheet" reflection traces so that there is an equal spacing of traces between fiducial marks, creating a normalized horizontal scale on all reflection profiles. When data are collected in steps or with a survey wheel, the

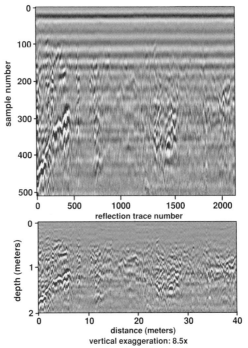

FIGURE 6.2
Raw Reflection Profile and Adjusted Profile. The upper profile contains only the raw reflections with no scale. They are measured in samples for the vertical dimension and reflection traces in the horizontal. The lower profile has been corrected for the time zero lag and adjusted in both the horizontal and vertical dimension to meters, with the vertical exaggeration noted.

placement of reflection traces in distance along transects is already normalized, but a horizontal and vertical scale must still be inserted to create the final product (figure 6.2).

During this process, it is always important to correct for the zero offset position that was set prior to data collection (figure 4.2). This is the offset in time between zero in the time window and the first reflection from the ground surface (often 1 or 2 nanoseconds). Its adjustment will place all reflections in the correct two-way travel time below the ground surface. If relative dielectric permittivity or the average velocity of the material through which radar energy is traveling is known from velocity tests, the arrival times of all reflections can then be adjusted to approximate depth in the ground. If velocities have not yet been determined, it is always good to leave the vertical scale in nanoseconds and adjust it later after velocity has been calculated.

If profiles were collected over uneven ground or exhibit other discontinuities due to velocity changes or surface vegetation differences, reflection traces can be easily stacked during this process. Stacking, which applies a running average and smoothes reflection data, will combine a certain number of traces (chosen at the discretion of the user) into one, in an arithmetic averaging process. This function is often performed during data collection in the field (chapter 4) but can also be performed at this time.

When data are collected over topographically complex areas, surface elevations must be compensated for so that all subsurface reflections can be adjusted to their correct positions in space (figure 3.8). If this is not done, the orientation and placement of subsurface features can never be known exactly, and often very strange interpretative results are created. Surface elevation correction of reflection profiles is a processing step that is a form of *static correction*, and it is often referred to as such in software commands. Besides topographic adjustments, static corrections can also adjust for changes caused by differences in antenna coupling or lateral velocity variations. When sites have a great deal of topographic relief—for instance, those conducted on the flanks of large mounds or hills and valley sides—there is no adequate way to perfectly adjust reflections in the ground (Lehmann and Green 2000). This is because the ray paths of radar waves in the ground are incredibly complex when antennas are tilted on a slope and the cone of energy transmission, and therefore the sources of reflections in the ground is always changing. There is no software yet developed that can adequately account for this type of transmission and reflection complexity, although some have attempted to do so.

REMOVE HORIZONTAL BANDING

The most common type of filtering that can be applied to any data set is the removal of horizontal banding that appears in most GPR reflection profiles (figure 6.3). Due to the "ringing" of some antennas, horizontal bands are recorded on most profiles (Leckebusch 2003; Shih and Doolittle 1984; Sternberg and McGill 1995). Often banding of this sort can be generated as "system noise," which is inherent to any GPR unit and from nearby radio and other electromagnetic frequency interference. Horizontal bands in profiles can also be the product of reflections from surface objects that were the same distance away from the antenna during acquisition, such as the person pulling the antennas or a towing vehicle. However they are produced, banding often obscures important subsurface reflections that would otherwise be visible on profiles.

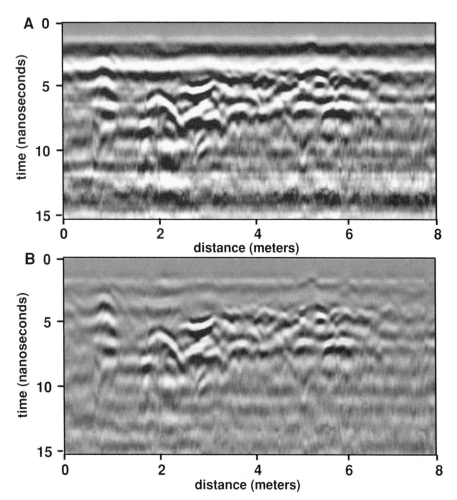

FIGURE 6.3
Background Removal Processing. The unprocessed reflection profile (A) has horizontal banding that obscures many important reflections. When this banding is removed (B), nonhorizontal reflections, especially in the upper 5 nanoseconds of the profile, become visible.

Most processing programs have the ability to remove these bands in a simple arithmetic process that sums all the amplitudes of reflections that were recorded at the same time along a reflection profile (or within a certain number of contiguous reflection traces) and divides by the number of traces summed. The resulting composite digital waveform, which is hopefully an average of all the background noise producing the horizontal banding, is then subtracted from the data set. These postfiltering background-removed profiles then will display only the nonhorizontal reflections, or those horizontal reflections that are short in length. Some processing programs can also remove a running average of any number of reflection traces desired in each reflection profile (Leckebusch 2003).

Great care must be taken when applying a background removing filter in areas where subsurface stratigraphy, or the features of interest, are horizontal, or nearly so. If this processing step were to be used on GPR profiles that have recorded horizontal reflections produced from horizontal buried interfaces, most, if not all, of those important reflections may be subtracted. Background removal should also only be applied on a digital reflection data set with a sufficient number of reflection traces. This is because background removal filters remove all the waves that occur at the same time, leaving only those that are more random. If too few reflection traces have been acquired in one profile over too short a distance, or if only a few traces are averaged to create the composite wave to be subtracted, the composite wave to be removed could be composed of all the important reflections in the profile (Leckebusch 2003). The averaging involved in removing the background would then remove most of the reflections from the ground, as well as the noise, leaving little reflection data of any sort.

Horizontal banding noise can also be removed with frequency filters, both high- and low-pass. Horizontal banding noise can also be caused by low-frequency interference that produces long-wavelength variations within reflection traces. It can be removed during data collection by applying the correct high-pass filters, as discussed in chapter 4, or, if the frequencies creating the noise can be isolated, removed during postacquisition filtering steps (Malagodi et al. 1996).

If all these processing steps will still not "clean up" reflection data, or if unusual conditions in the subsurface or a good deal of electromagnetic interference are encountered, additional steps may be necessary. The following techniques are not commonly used by most archaeologists, but in certain situations they can greatly improve data quality and clarity. In some drastic cases, a very large percentage of the original raw data must be removed in order to enhance the remaining reflections and produce quality results (Conyers and Cameron 1998).

REMOVAL OF HIGH-FREQUENCY NOISE

This process can get quite complex as there are a number of processing parameters that can be adjusted in order to enhance data quality. If there is a great deal of high-frequency interference from cell phones, pagers, radio transmitters, or other devices nearby, reflection profiles may contain so much "snow" and "banding" as to be uninterpretable. When this occurs. background removal, which is usually the first and simplest processing step, may still yield very poor reflection profiles. A number of filters in some processing programs attempt to remove the interfering frequencies with either *infinite impulse response* (IIR) or *finite impulse response* (FIR) filters. Both filtering processes remove certain frequencies with the difference being the number of reflection traces used in averaging and the type of averaging and interpolation algorithms used in the filtering process. In all cases, what is being done is a sophisticated method to remove specific high (and sometimes low) frequencies that might be contributing noise.

There are many experimental filtering techniques of this sort, originally developed for seismic data processing by the petroleum exploration industry, that have been applied to GPR reflection data (Lehmann et al. 1996; Maijala 1992; Milligan and Atkin 1993; Yu et al. 1996). Care must be taken when using these processing steps because there are some important differences between radar and seismic reflection data. The wide aperture of radar antennas, which transmit and receive reflections back from a large subsurface area in a cone of transmission, make any automatic application of seismic techniques sometimes difficult as their wave transmission properties are much different. Another difference is that radar energy slows down during its propagation in the ground, while seismic waves will increase their velocity with depth (Leckebusch 2003).

F-k filtering is one such technique borrowed from seismic processing where reflections recorded in time are transformed into frequency data using statistical transform programs (Maijala 1992). The outcome of this processing procedure is that high-angle reflections (possibly also point source reflection hyperbolas), which may be obscuring important horizontal data, are removed. This seismic data-processing technique has been misapplied to some archaeological GPR data and should only be used with caution.

REMOVAL OF MULTIPLE REFLECTIONS

Multiple reflections (often just called *multiples*) are caused when radar energy reflects back and forth between a buried object and the ground surface, or between subsurface layers. When each of the reflections from these objects or interfaces are received at the surface antenna and recorded, they are displayed in profile as repetitive horizontal reflections (figure 6.4), all of which are an equal distance apart (as measured in time). Often multiples of this sort can be confused with "real" reflections that might have been created from multiple stacked layers in the

POSTACQUISITION DATA PROCESSING

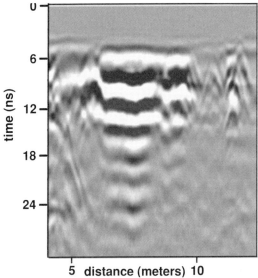

FIGURE 6.4
Multiple Reflections. These stacked multiples were generated from a horizontal surface with a high coefficient of reflectivity. Radar energy was reflected from the buried interface back to the ground surface, and then back into the ground, to be rereflected numerous times from both interfaces, creating the stacked reflections.

ground. Usually they can be differentiated because multiples are spaced at an equal interval in time as they are stacked on top of each other within reflection profiles. Sometimes multiples are also created by "ringing antennas" that are located within each other's near-field, and therefore reverberate, creating the same effect.

These unwanted multiple reflections often obscure important reflections that might have been recorded at the same time in the reflection traces, and if so, they must be removed. This process is called *deconvolution*, or more precisely *predictive deconvolution*, because the method attempts to predict the shape of the transmitted radar pulse from the surface antenna and how it will change as it moves in the ground. This processing technique is another step that has been modified for GPR data from seismic studies (Fisher et al. 1994; LaFleche et al. 1991; Maijala 1992; Malagodi et al. 1996; Neves et al. 1996; Rees and Glover 1992; Todoeschuck et al. 1992). It is based on the theory that as a radar pulse is transmitted into the ground portions of the electromagnetic wave will change form, or convolve. The purpose of this filter is to remove the portion of the recorded waves that have convolved during transmission in the ground. Deconvolution processing restores the reflected waves in a profile to their original pattern and presents the data with a "different look." The deconvolution technique can be used to identify and then remove multiple reflections, if they can be accurately predicted.

One of the problems with deconvolution processing is that restoring reflected waves to their original forms is mostly educated guesswork. It is usually difficult to determine how the original transmitted waves were shaped and any deconvolution process may be modifying the data in unreal ways. Although many GPR experts claim to understand what radar waves do in the ground, much about this processing technique still remains obscure. If deconvolution processing of radar data can be improved, it may yield important clues to understanding how the physical properties of the ground modify transmitted electromagnetic waves and help in all types of GPR data interpretation. Most attempted applications of this processing step have so far proved to be unsatisfactory (Maijala 1992; Rees and Glover 1992) or, at worst, have removed important reflections unintentionally.

MIGRATION

Standard GPR systems portray a distorted image of subsurface stratigraphy and features in the ground. This is caused both by the wide beam of radar propagation producing multiple ray paths as well as changes in the velocity, and therefore refraction of the transmitted beam with depth. Migration is a two-dimensional imaging process that has been used with success to eliminate some of these distortions caused in all reflection data collection procedures (Beres et al. 1999; Fisher et al. 1992; Fisher et al. 1994; Grasmueck 1994; Malagodi et al. 1996; Milligan and Atkin 1993; Young and Jingsheng 1994). The distortions that can be most readily corrected by migration are caused by the wide transmission beam of radar antennas that generate reflections from point sources, appearing as hyperbolas (figure 3.17). Before reflection profile data can be processed into two-dimensional maps and three-dimensional images, these hyperbolas often must be removed (Conyers et al. 2002).

Steeply dipping surfaces will also diffract radar energy during its transmission to and from a reflecting surface (Jol and Bristow 2003). Longer travel times that result from this diffraction will place reflections at incorrect depths or locations in the subsurface, distorting the size and geometry of some subsurface beds or features. The migration process can be used to spatially adjust for these kinds of distorted or hyperbolic reflections and "collapse" them back to the point of origin (figure 6.5).

The easiest migration method for hyperbola removal is accomplished by summing all the reflections along a hyperbola's arms and placing the resulting average at its apex. This must usually be done manually for all hyperbolas in a profile, which usually necessitates first identifying all of them, which can be a tedious process. A more sophisticated process called the *Kirchoff method* (Geophysical Survey Systems 2000) calculates the angle of incidence and the distance the reflecting feature is below the surface. It then applies velocities or velocity profiles (which often must be assumed) to the data and corrects the placement of reflections, also effectively collapsing all hyperbola arms back to their apexes. The same processing procedure can be used for steeply dipping beds, where the same type of geometric corrections can be made. There are three other migration algorithms used for migration called *Stolt*, *phase-shift*, and *finite-difference*. Tests conducted by Leckebusch (2003) showed no difference between them or with the Kirchoff method.

Migration is becoming a standard technique in most GPR processing programs, but it can be very time-consuming and also potentially distort many reflections incorrectly. It should therefore be employed for very specific types of data analysis, and only after one is sure of the origin of the distorted reflections, the velocity of materials in the ground, and the geometry of wave travel paths in the subsurface.

In addition to the more standard two-dimensional reflection profile migration, three-dimensional migration is possible (Grasmueck 1996; Leckebusch and Peikert 2001; Shrugge and Artman 2004). This method will collapse not only hyperbolic reflections that are generated "in-line" (along the direction antennas are moved in transects) but also reflections that are received from out of the plane of the antenna transect (from the sides).

In many cases, it may be better to view radar reflection profiles in an unmigrated format first because point source hyperbolas are many times capable of identifying subsurface anomalies that represent archaeological features of interest. If all the hyperbolas were collapsed back to the point of origin (assuming that migration processing techniques are being properly applied), point sources are often more difficult to recognize, possibly causing some important buried features or objects to go unrecognized when viewed in profile. The opposite could also be true if important reflections (other than those creating the hyperbolas) were generated within the axes of the point source hyperbolas. In this case, the migration of hyperbolas back to their point of origin might allow important but otherwise obscured reflections to become visible in reflection profiles.

INCREASE THE VISIBILITY OF SUBTLE REFLECTIONS

Low-amplitude reflections can always be enhanced so as to become visible in reflection profiles by increasing the range gains at certain recording times (depths

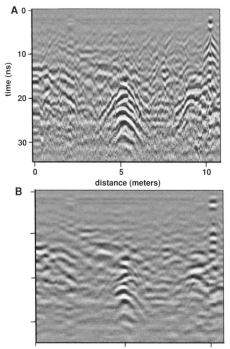

FIGURE 6.5
Reflection Migration. Reflection hyperbolas generated from the tops and sides of buried stone walls at Petra, Jordan (A), are migrated back to their origins (B), cleaning up the profile and leaving only the hyperbola apexes.

in the ground). Many processing programs have the ability to move a mouse along a gain curve, increasing the amplitudes at specified times for all reflection traces in a profile (figure 4.1). When the amplitudes of all reflections in a desired time window are increased, background removal filters must then be applied as a second processing step to remove resulting horizontal banding, which is also increased. This process can be repeated until otherwise invisible reflections are visible. The risk in this series of processing methods, as with all postacquisition processing, is that each step filters and potentially distorts the initial waveforms, making the final product potentially so distorted as to be unrecognizable.

Reflections generated from very subtle changes in the ground can also be enhanced with a processing step called a *Hilbert transform* (Sensors and Software 1999; Todoeschuck et al. 1992; Turner 1992; Yilmaz 1987). This step transforms the reflection amplitudes and their geometry in the ground into spatially distinct frequency and phase information. The phase of reflections (whether they are positive or negative deflections in the waveform from a mean) is often indicative of important changes in relative dielectric permittivity of materials at a reflecting interface. For instance, the phase of a reflected wave will often be different if generated at a void space (with an RDP of 1) than it would be for a harder or denser object (that might have a much greater RDP than the surrounding material). The phase change in amplitudes along a profile can therefore potentially tell the interpreter what type of material is generating the reflection. The same is true for frequency changes along a reflection of interest, as they indicate how the ground is "filtering" the radar energy during transmission and as it is reflected back to the surface.

DATA PROCESSING CONCLUSIONS

As a matter of course, all reflection profiles should be processed to normalize distance both horizontally and vertically. Background removal is also a very simple and important technique that should be attempted to determine if it can improve the visibility of important subsurface reflections. All but these simplest GPR data-processing steps can potentially distort reflection data out of recognition and should be used with caution. While very noisy or distorted data can often be corrected and filtered so as to be more usable, there is always a risk in "overprocessing." In the seismic processing business, some people have devoted their lives to postacquisition data processing and can often work wonders with poor-quality data. But they also make jokes among themselves about their ability to modify data to such a degree that they can produce almost any result desired by a client. The standard joke in the seismic data-processing laboratory is that if the reflection data do not produce the desired outcomes, just send them back to the data-processing experts, as they can produce whatever results one would like!

Although it is hoped that one would never become this intellectually confused within the data-processing web, the potential exists with the abundance of processing steps discussed here, few of which are really understood in their complexity by many GPR professionals, let alone the typical archeological user.

7
Interpretation of GPR Data

When GPR data were first applied to archaeology in the 1970s, most interpretations were made by viewing raw reflection profiles and searching for "anomalies" that might have been produced by reflections from buried archaeological features (Bevan 1977; Sheets et al. 1985). Often reflection profiles were printed out and analyzed on paper as they were collected in the field or viewed and interpreted on a video monitor. These interpretation methods were at best crude and inaccurate, but they were about all that could be done with analog reflection data that were not stored on some kind of digital medium for later processing.

Unfortunately, this kind of interpretation is still practiced by many GPR users, but usually only when data are being acquired in the field and are immediately viewable on the computer screen as they are collected. Most GPR users that still employ these kinds of on-the-fly interpretations are aware of the pitfalls that can be encountered when trying to make interpretations in this manner. All should now be aware that field records should be processed to remove noise, scales must be corrected in either the vertical or horizontal directions and that picking out anomalies on the screen that may or not have any meaning to the questions being asked (assuming they can be recognized!) is fraught with problems.

With the advent of digital data filtering and sophisticated data processing, in the late 1980s, profiles could finally be interpreted in a more refined and scientific manner than was previously possible (Conyers 1995; Goodman 1994). However, until the advent of three-dimensional amplitude analyses in the mid-1990s, most GPR data interpretation consisted of analyzing reflections in profiles that "looked like" buried features of interest and correlating them to other reflections in adjoining profiles in a method that has been termed "wiggle picking" by some interpreters. Using this technique, buried features can still often be identified, their depth determined, and their spatial placement within a grid mapped, but the process is laborious and has many problems. Using the visual interpretation method, interpreters are often unsure which reflections were generated from features of interest and which might have been produced from geological "background," and therefore the origin of many mapped anomalies are often little more than educated guesses. Unfortunately, this kind of data interpretation is still practiced for some GPR studies as many practitioners have not applied many of the more sophisticated interpretation methods discussed here, such as synthetic modeling, amplitude time slicing, and three-dimensional rendering. But even when these more sophisticated interpretation methods are applied, it is almost always necessary to still use some kind of "wiggle picking" analysis. This is usually done so that the maps and images being produced can be correlated with features identified in profiles that are of archaeological interest. All GPR studies should therefore include some components of two-dimensional reflection profile analysis, modeling, and three-dimensional mapping, as discussed in this chapter.

SYNTHETIC GPR MODELS

Synthetic modeling of two-dimensional reflection profiles using a computer was developed in an attempt to model buried objects, stratigraphy, and important reflection surfaces (Goodman 1994). Modeling can provide the interpreter with an idea of what real-world GPR reflection data "should look like" and will allow more accurate interpretation of GPR reflection profiles once they are processed (Conyers and Goodman 1997; Goodman 1994; Goodman and Nishimura 1993). It can also allow the interpreter to construct a model of known stratigraphy and archaeological features prior to going to the field in order to determine if a GPR survey will be capable of delineating the buried materials of interest. Once models are constructed on the computer, they can be quickly modified for different-frequency antennas to determine the optimum equipment to take to the field for the depth necessary to resolve the features of interest. After GPR data have been acquired and are processed into reflection profiles, models can then be readjusted to more accurately represent known field conditions as a guide in interpretation. When used in this way, they are a great benefit in making interpretations, especially in determining the origins of reflections visible in GPR profiles.

Computer-simulated radar profiles are generated by tracing the theoretical paths of radar waves during transmission through various media with specific relative dielectric permittivities, electrical conductivities, and magnetic permeabilities. All possible reflections from modeled interfaces in the ground are taken into account (Goodman 1994). The two-dimensional geometry of the subsurface stratigraphy and archaeological features are programmed into the model to generate as close to a real-life case as possible.

As is often the case, two-dimensional reflection profiles can look significantly different from how the buried structures would appear in cross section if viewed in the wall of a trench. Most important, they are not at all like those most of us are used to seeing from other more common computer-enhanced images such as those from X rays or computer tomography (CT) scans in medical technology. Once synthetic models are studied and the generation of reflections can be compared to the modeled archaeological features, they can also be an excellent learning tool to help understand how radar energy is propagating and reflecting in the ground.

Synthetic models of important components of an archaeological site can only be produced if some prior information about subsurface conditions is available. The electrical and magnetic characteristics of sediment and soil conditions must often be estimated as well as the geometry of overburden units and the composition of the buried archaeological features of interest. These data are then put into the computer in a two-dimensional model that is a simplification of a vertical slice through the ground. The computer will use this information to predict reflectivity coefficients (equation 3.2) encountered at various interfaces, energy attenuation with depth and within each unit, the velocity of radar energy in different layers, and the amplitude of reflections received back at the surface (Goodman 1994). After the model is generated, resulting reflections are plotted in two dimensions in the same fashion as standard GPR reflection profiles, and its horizontal and vertical scales can be adjusted, just as with real reflection profiles. After analysis of any model, input parameters can then be changed and the simulation rerun until a reasonable match between the real and synthetic reflection profiles is obtained. This iterative process of parameter input and comparison to real-world data is referred to as *forward modeling* in geophysical prospecting (Powers and Olhoeft 1994, 1995; Zeng et al. 1995) and is a powerful interpretation tool.

Creating a Synthetic Computer Model

To generate a synthetic reflection profile, large numbers of potential radar wave paths through the two-dimensional model are calculated on a computer, which approximate the paths that radar waves would take in the ground. This type of two-dimensional modeling is well known in seismic processing and is often referred to as *ray tracing* in geophysical terminology (Cai and McMechan 1994; Goodman 1994; Leckenbusch and Peikert 2001). Unlike seismic modeling, GPR models must also take into account the conical-shaped transmission patterns of the surface transmitting antennas (Goodman 1994). Ray path models in GPR are based on algorithms that take into account a number of complex equations concerning conductive-dissipative electromagnetic propagation theory (Jackson 1977).

Some possible wave paths that must be considered when creating a synthetic computer reflection profile are shown for the example in figure 7.1. These, and many other possible ray paths similar to them, must be programmed into the computer to generate an accurate model of what radar energy will do in the ground. In the figure 7.1 example, there are four different media in which radar waves can travel through and be reflected from: air, unit 1, unit 2, and unit 3. A synthetic reflection profile that would be produced for this simple four-layer model would therefore need to take into account waves that are reflected off each subsurface interface, but also ray paths that are partially reflected off some interfaces and partially transmitted and refracted across others. At least six separate ray travel and reflection pathways would have to be incorporated into the model (see figure 7.1). In this simple example of possible wave paths, some of the radar waves will reflect off the interface between units 1 and 2 and will be recorded back at the surface antenna. Others will be transmitted through this interface and refracted, only to be reflected off the interface between unit 2 and unit 3. In the real world, only a portion of the energy from each energy pulse would be reflected and refracted at each boundary, but for simplicity the model is set so that each radar ray has one distinct path. For instance, one wave will be reflected off the interface between units 1 and 2, traveling directly back to the surface where it is recorded at the receiving antenna. Another wave travels the same path, but instead of being recorded at the receiving antenna it is rereflected back into the ground from the air–unit 1 interface. The waves of these multiple reflections will then travel back into the ground to be rereflected back to the surface at the same interface. A similar ray path with multiple reflections is also simulated at the interface between unit 2 and unit 3. There will also be ray paths simulated that take what are called "dogleg" paths, traveling in complicated multiple reflection paths off of all the modeled interfaces prior to finally arriving back at the surface. These, and other paths within the cone of illumination of the transmitting antenna, are simulated on the computer many thousands of times to arrive at a composite for all the reflected waves that arrive back at the surface antenna. The computer then moves the antenna along the modeled ground surface, and the process is repeated for many surface antenna locations along the programmed transect. The computer predicts the time at which the modeled waves return to the receiving antenna on the surface and records their resulting amplitudes.

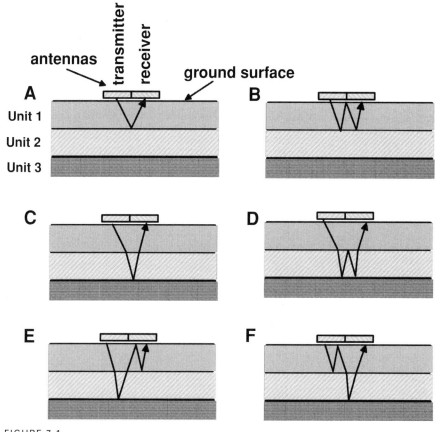

FIGURE 7.1
Possible Ray Paths in Stratified Material. To model potential radar reflections in the ground, many radar wave pathways must be considered. The simplest path is from the surface antenna to a buried interface and back to the surface antenna (A). Multiple reflections occur in (B). Energy can also be refracted and then reflected (C) or any combination of reflection and refraction (D, E, F).

To compute the synthetic profile, all the modeled waves that are returned to the receiving antenna from each of the many hundreds of paths are totaled at each antenna location over the profile. The time at which each is received and the resulting amplitudes are also recorded, and the simulated reflection traces are plotted in a standard two-dimensional profile. The number of reflections or transmissions that occurred at each interface can be adjusted by the modeler. Using the RDP and electrical conductivity data that are input for each simulated stratigraphic unit, the computer calculates the reflection coefficients at each interface and the amount of radar energy that is reflected or transmitted. Attenuation along the path of each modeled wave is also computed from the relative dielectric permittivity and electrical conductivity estimates that are programmed into the computer program. In the real world, there are an infinite number of possible wave travel paths and resulting amplitudes, but if the model is constructed without too much complexity, large numbers of possible ray paths will generate a coherent and usable synthetic model (Conyers and Goodman 1997; Goodman 1994).

Details about the electromagnetic theory involved and the mathematical description of all steps in the model is given in Goodman (1994). In real field conditions, numerous out-of-plane reflections must also be taken into account (Carcione 1996; Grasmueck 1996; Leckebusch 2003). Much more complicated three-dimensional modeling techniques have recently been developed for seismic reflection data used in petroleum exploration, which are just starting to be applied to GPR studies (Lehmann et al. 2000).

Synthetic Modeling Applications

In areas where buried features have considerable geometric variation within a short distance, GPR reflection profiles and the models of these buried features can be very complex. An example of a synthetic reflection profile for a buried V-shaped trench with steep sides and two different subsurface layers is shown in figure 7.2. The model predicts that some radar waves will have a single reflection off the subsurface interface. Others will have multiple reflections in the ground from within the trench, and these are also taken into account. The direct reflected waves (those with only one reflection off the interface) show the outline of the trench, with less amplitude on its edges due to ray scattering that occurs from its sloping sides. Because the antenna will "see" the far wall of the trench before encountering it, due to the conical transmission pattern from the surface antenna, reflections from the trench side will be recorded (in measured time) at a depth greater than the actual trench bottom. The same is true as the antenna moves

INTERPRETATION OF GPR DATA

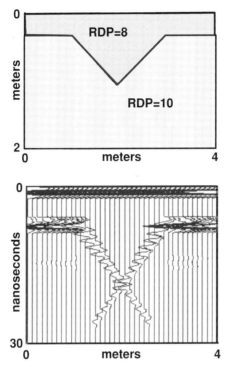

FIGURE 7.2
A V-Trench Synthetic Model. A synthetic model where radar wave paths travel to and from a V-shaped trench. Waves are reflected multiple times within the trench, before energy reaches the receiving antenna. The result in this model shows a "bow tie" pattern in the modeled reflection profile.

away from the trench and reflections are received from its opposite side. The net result of these reflections, as the antennas are moved along the ground surface, causes what is referred to as a "bow tie" effect when the model is viewed in profile. Other multiple reflection features are also created as radar waves are reflected three or more times within the trench but are recorded as very low-amplitude waves due to progressive energy attenuation during transmission through the ground.

In the V-trench case (figure 7.2), if the synthetic model were not constructed and analyzed, the reflections beneath a trench of this sort might be mistaken for a point source hyperbola derived from something possibly buried in the bottom of the trench, or perhaps some other type of buried feature below the trench. The model, however, demonstrates that these reflections in the profile are the result of multiple reflections from inside the trench and do not represent a "real" feature. When similar trenches are seen in real reflection profiles at a site (such as the one in figure 7.3), they can be easily identified in profiles because the model has simulated and predicted their shape in advance.

A model that illustrates how layered materials of varying thickness, with differing RDP, produce reflections is shown in figure 7.4. In this model, the flat interface between units 2 and 3 represents a buried living surface and is the archaeological interface of interest. The complexity in this model arises from the differences in RDP and thickness changes in the two overlying units.

In figure 7.4, the thick section of material with a high RDP in unit 2 will slow the radar energy as it travels vertically from the ground surface, to the interface of interest, and back to the ground surface. The thinner section of this high-RDP material in the middle of the model illustrates how radar energy will travel from the ground surface to the interface of interest, and then back to the ground surface in a shorter amount of time than at the edges because it is traveling for less distance in lower-velocity material. The resulting reflections generated from the interface of interest at the top of unit 3, when plotted in two-way travel time, will therefore be distorted due to these velocity and thickness differences in the overlying unit. Under the area where a thinner section of low-velocity material (higher RDP) is modeled, the lower interface will

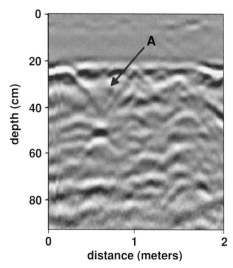

FIGURE 7.3
A V-Trench Discovered in a Reflection Profile. This V-shaped trench generated a classic bow tie reflection (A) like that modeled in figure 7.2. This profile was collected in a highly disturbed historic mining area near Blackhawk, Colorado.

INTERPRETATION OF GPR DATA 141

appear to bow upward. This upward and downward bowing of the lower interface, caused only by differences in the velocity and thickness of the overlying material, creates the illusion of an undulating surface of the interface between units 2 and 3, referred to as a *velocity pull-up* in one area and a *pull-down* in the other.

A velocity pull-up similar to the model in figure 7.4 would also be noticeable in the field below a large buried void space. The increase in radar wave velocity as waves are traveling within the void would create an artificial upward bowing of reflections that were generated from materials below it. A pull-down could also occur where localized conditions create a decrease in radar wave velocity, possibly due to abrupt stratigraphic or archaeological change laterally along a feature or a possible change in overlying soil conditions. A fluctuating water table, or changes in the water saturation of buried units located above the water table, can also slow radar waves and distort underlying reflections due to localized variations in velocity. This phenomenon was noticed in a sand dune area where a pipe, which was known to be horizontal, appeared in a reflection profile

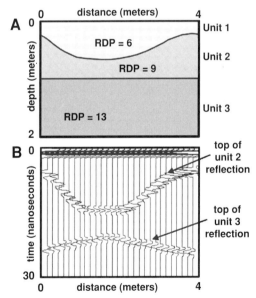

FIGURE 7.4
A Three-Layer Velocity Model. Three layers of material are modeled with a high RDP unit (unit 2) sandwiched between units 1 and 3 are shown in A. Radar energy travels for a shorter distance through the lower-velocity material of unit 2 in the middle of the model, creating a velocity "pull-up" of the reflection at the top of unit 3 and "pull-downs" on either side.

to undulate (figure 7.5). Its pull up and pull down is only a function of differences in RDP in the overlying sand dune beds, much like demonstrated in the model in figure 7.4.

The model in figure 7.4 and the undulating pipe, which is actually horizontal (figure 7.5), demonstrate one of many pitfalls that interpreters of GPR data can encounter when there are large changes in velocity within the ground overlying buried materials of interest. In addition to these problems, because of complex refraction and transmission within subsurface layers, radar energy may not be transmitted at all through some types of materials because of total energy attenuation or complete reflection. These "unilluminated" regions that occur below these types of materials are also referred to as *shadow zones* (Goodman 1994). They are areas where no reflections occur even though features that could potentially reflect energy are present. Often these and many other "non-intuitive" reflection and transmission scenarios would not be recognized without first generating synthetic models of real-world conditions and studying their outcomes.

Synthetic Models Compared to GPR Profiles

When a relatively simple reflection profile of a buried pit house was processed and analyzed, its floor and possible subfloor features were immediately visible (figure 7.6). The house floor was then cored to confirm its presence and also to evaluate the properties of its compacted and partially baked floor and the sedi-

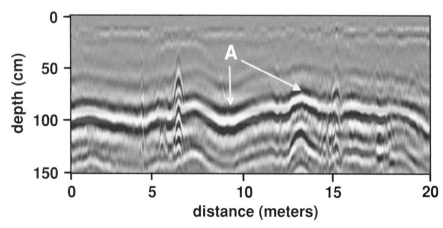

FIGURE 7.5
Vertical Distortion Due to Near-Surface Velocity Differences. A horizontal pipe appears to undulate because of velocity changes in the overlying sand dune beds. At points marked (A), there are both pull-ups and pull-downs.

ment covering it (Conyers and Cameron 1998). Upon further analyses of each profile crossing the floor feature, additional subtle features were visible on or below the floor, whose origin was unknown. One of the most distinct was what appeared to be a break in the floor surface that was hypothesized to be an entrance to a subfloor cistern, which is common in pit houses in the area of the American Southwest where these features were found (figure 7.6).

A synthetic model was constructed of the floor with a subfloor storage cistern, to determine if this type of cultural feature could be producing what was seen in the reflection profile (figure 7.7). Information obtained from the core was used to determine sediment properties used in the model. The synthetic model indicated that a cistern of the sort modeled would probably not be visible in reflection profiles, except for perhaps a very subtle reflection at its base. Its sides would probably be invisible because of their vertical faces, and all reflections below the highly reflective floor would also be very weak, as most of the radar energy would be reflected back to the surface producing a shadow below it. The model supports the presence of a possible subfloor feature in the real reflection profile as it shows a

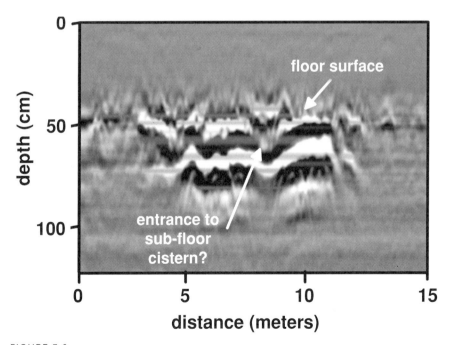

FIGURE 7.6
A Pit House Floor with Possible Subfloor Feature. The gap in the floor reflection may be the entrance to a subfloor cistern, which is not otherwise visible.

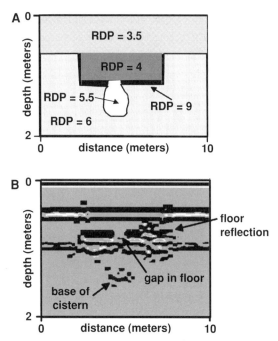

FIGURE 7.7
Synthetic Model of a Pit House Floor. This model indicates that the vertical walls of the subfloor cistern would not be visible in a reflection profile because little energy is reflected from them. The base of the large storage cistern is barely visible, as most of the radar energy is reflected back to the ground surface from the floor, producing a shadow below. The only clue to the subfloor feature is a gap in the floor reflection, similar to what was visible in the reflection profile in figure 7.6.

break in the horizontal floor reflection, which is just like that visible in the actual reflection profile crossing the buried floor (figure 7.6). Without some confirmation from the synthetic model, a feature as subtle as this would probably go unnoticed by most interpreters.

INTERPRETING AND MAPPING MANY GPR REFLECTION PROFILES IN A GRID

In areas where well-defined GPR reflections are recorded in profiles, detailed subsurface mapping of the location of buried surfaces and archaeological features (if they can be identified) is possible. Prior to making maps from individual profiles, the genesis of the reflections must be understood so that the maps are meaningful, by visual interpretation, analysis of two-dimensional synthetic models, or both. Once the reflections in profiles are differentiated and the origin of important reflections are defined, they can be manually, or with the aid of a computer, correlated from profile to profile within a grid and between grids in an area. If velocity and surface topographic corrections have been made, accurate depth maps that define buried topography, or anthropogenic modifications to that landscape, can then be constructed (Conyers 1995; Conyers and Goodman 1997: 137; Imai et al. 1987; Milligan and Atkin 1993).

Manual profile interpretation of this sort can be very time-consuming and always relies on interpretive experience. The origin of reflections and their importance to the questions at hand must also be determined in advance or a great deal of energy might go into identifying and interpreting reflections that may not answer any archaeological questions. For this reason, velocity analysis, correlation of reflections to known materials in excavations, and two-dimensional modeling (or all the above) should be performed in advance of any manual profile interpretation.

Example of Buried Landscape Reconstruction from Interpreting Reflection Profiles

A standard manual GPR correlation and paleotopographic mapping technique was used to define and then map a buried sixth-century living surface at the Ceren Site in El Salvador (Conyers 1995). At this site, the ancient living surface is now covered by between 2 and 6 meters of volcanic tephra, which preserves the ancient landscape, architecture, and artifacts (Sheets 1992). The buried surface was initially defined in GPR reflection profiles using both velocity analysis and synthetic computer modeling (Conyers and Goodman 1997). This important reflection that correlates to the buried living surface was then correlated within all reflection profiles, and its subsurface attitude was manually mapped.

The top of the buried living surface is the most important interface to map with GPR at this site because it was the ground surface that was built on, farmed, and modified prior to burial by volcanic ash (Conyers 1995). During interpretation, each profile in the grid was first processed to remove background noise, adjusted for surface topographic changes when necessary, and frequency filtered. Velocity analyses including direct-wave studies and laboratory measurements of sediment samples were conducted to arrive at RDPs for each unit (Conyers and Lucius 1996). Two-dimensional models were produced for many of the architectural features that were known to have been constructed on the living surface so that they could be identified in profiles (Conyers 1995). Velocity information was then used to correct radar travel times to depth, and the final two-dimensional profiles were printed on paper for visual analysis. The reflection generated at the buried living surface could be identified through a long and laborious process of stratigraphic correlations within and between all reflection profiles. This was done by hand-coloring both the buried living surface reflection and all structures built on it on each paper profile. The elevations of all identified reflections were then measured with hand calipers every meter along more than 12,800 linear meters of paper reflection profiles.

Distinctive point source reflections with apexes above the clay floors of buried structures were visible in some profiles. It is probable that these point source reflections are the record of reflections that occurred from the tops or sides of standing columns or walls, as predicted in previously constructed two-dimensional synthetic models. These hyperbolic reflections and also some distinct reflections from horizontal clay floors were used to map all visible buried structures (figure 7.8).

This manual interpretation process took about two months, as the critical reflections in each profile had to be visually correlated profile by profile throughout nine different adjoining and sometimes overlapping GPR grids and "tied" at each perpendicular transect intersection to assure consistency. Subsurface elevations along each profile were then plotted on a grid map and contoured both by hand and using a computer mapping program to reveal the three-dimensional topography of the buried living surface (plate 1). The final maps were produced from more than 475 reflection profiles covering a surface area of more than 4 hectares. This time-consuming interpretation process ultimately produced good subsurface maps that were an accurate guide for the placement of archaeological excavations that could test features of interest.

The topographic, stratigraphic, and archaeological complexity at this site necessitated this type of manual data interpretation, which, if it were conducted elsewhere, would have to be budgeted for, in both time and money. Although some computer-processing programs claim to be able to automatically correlate reflections from profile to profile within grids, the stratigraphic complexity usually encountered at most archaeological sites makes this type of "automatic" interpretation tenuous at best. When clients or archaeological colleagues hear how time-consuming this type of GPR data processing can become, they are usually either totally enthralled with the amazing detail that can result or discouraged to think about what it might cost in time and money to reach a viable conclusion.

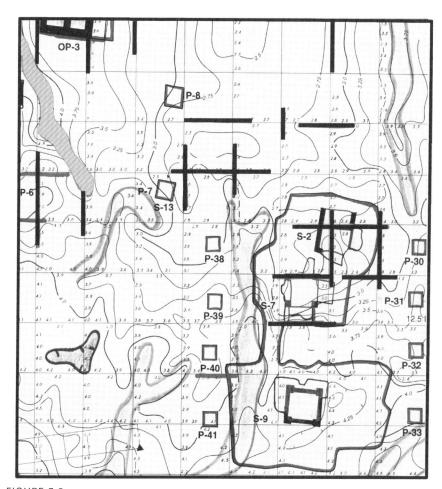

FIGURE 7.8
Hand Mapping of Reflection Profiles in a Grid. Many reflection profiles in a grid were manually interpreted, and those of importance were noted to produce a map of the buried landscape and cultural features built on it. This subsurface map of a portion of the Ceren site in El Salvador shows many water channels and house platforms.

In contrast to the above laborious processing method, a more recent GPR survey at a very complex buried site at Petra in Jordan covering an area of more than 80 × 50 meters was acquired in two days and produced a data set that generated usable and important results in about two additional days of data processing, using the amplitude analyses techniques discussed later in this chapter (Conyers et al. 2002). The reflection data at Petra, however, are so complex and potentially interesting that they still deserve months of manual interpretation in order to derive all the information that is contained in them. It is doubtful this will ever occur. This typical scenario of "GPR data overload" can be both a blessing and a curse, depending on the processing and interpretation time that can be budgeted for, the questions that need to be answered, and the resources available. Sometimes difficult decisions must be made as to how much information can or must be gathered from each grid of GPR data collected. Often only a small fraction of the subsurface information that GPR data can potentially yield is extracted from each survey, leaving many hundreds of CDs and other storage media filled with potentially important data just waiting for hypotheses to be generated and tested, time for interpretation, and usually funds to pay for it. When time is not available for the type of manual profile analysis conducted at the Ceren site, some combination of manual and visual processing can still be combined with the amplitude analysis techniques discussed here in order to produce very usable and accurate maps and images, which can often be produced and interpreted in a matter of hours.

AMPLITUDE ANALYSIS IN SLICE MAPS

The primary goal of most GPR surveys in archaeology is to identify the size, shape, depth, and location of buried cultural remains. The most straightforward way to accomplish this is by identifying important reflections, correlating them within two-dimensional reflection profiles as discussed earlier, and then mapping them, but this can be very time-consuming and sometimes inaccurate. A more sophisticated type of GPR data manipulation is amplitude slice map analysis that creates maps of reflected wave amplitude differences both spatially and with depth in a grid (Conyers and Goodman 1997; Goodman 1996; Goodman et al. 1998). The result can be a series of maps that illustrate in three dimensions the location of reflection anomalies, derived from a computer analysis of many two-dimensional profiles. What is being mapped in this method are the amplitudes of reflected waves (and their recorded depth in the ground), which are proxy measurements of the differences in materials at buried interfaces that reflect the radar energy. This method of data processing can only be accomplished by a computer using GPR reflection data that are stored digitally.

Raw reflection data collected in most GPR studies are nothing more than a collection of many hundreds of thousands of individual reflection traces along two-dimensional transects within a grid. Each of those reflection traces contains a series of waves that vary in amplitude depending on the intensity of energy reflection that occurred at buried interfaces. When these traces are plotted sequentially in standard two-dimensional profiles, the specific amplitudes within individual waves that contain important reflection information are usually difficult to visualize, and interpretation can be arduous. Also, in the past when raw, unprocessed GPR reflection profiles had no discernible reflections or recognizable anomalies of any sort, the survey was usually declared a failure, and little if any further interpretation was conducted. Only with the advent of more powerful computers and sophisticated software programs in the mid-1990s that could manipulate large sets of digital data was important subsurface information in the form of amplitude changes at various depths within a grid extracted from tens or sometimes hundreds of individual profiles possible. This method can produce three-dimensional packages of amplitudes, which can be analyzed in bulk, quickly producing maps and images of important high (or sometimes low) amplitudes that were generated from buried archaeological or related geological features of interest (plate 2).

An analysis of the spatial distribution of the amplitudes of reflected waves is important because it is an indicator of subsurface changes in lithology or other physical properties of buried materials. The greater the amplitude of a reflected wave, the greater difference in physical and chemical properties of materials at a buried interface reflecting that radar energy. Areas of low-amplitude reflections usually indicate uniform matrix material or soils, while those of higher amplitude denote areas of high subsurface contrast such as buried archaeological features, voids, or important stratigraphic changes. In order to be correctly interpreted, amplitude differences must often be analyzed in horizontal slices that examine only changes within specific layers in the ground.

Each amplitude slice of a certain thickness is comparable to an arbitrary archaeological excavation level, except using GPR data each level consists of a spatial representation of reflected wave amplitudes instead of sediment, soil, or feature changes and associated artifacts. In plate 2, stone foundations of buried buildings, which contrast with a wind-blown sand matrix, produce high amplitudes, which are readily apparent in amplitude slice maps. This example, from the Petra site, in Jordan (Conyers et al. 2002) had excellent velocity contrasts at the interfaces between the architectural stone and the surrounding sand, producing very high-amplitude reflections at their interfaces, which could be easily interpreted.

When spectacular GPR maps, such as those shown in plate 2, are used as an example of what GPR can do there is always the hope that similar data processing and image production at less definitive sites will produce similar results. Often this is not the case. Although amplitude slicing will produce excellent maps in some cases, or marginally usable maps in others, using only this technique for interpretation is usually not enough. It is a mistake to produce only generic slice maps in the hope that what one is searching for will immediately appear in the resulting images. Slice maps cannot be made automatically, like having photographs developed, but must be constructed thoughtfully and adjusted for various site parameters such as depth of interest, dimension and orientation of buried features, and the nature of the surrounding matrix. Their production therefore requires some prior knowledge of site conditions and often a detailed analysis of individual reflection profiles that have been used to produce the maps, all of which have hopefully been processed and filtered in some way to improve reflection data quality. Raw reflection data should also never be given to someone to slice who is not familiar with their collection, as they will have limited knowledge about site conditions and may therefore not be able to determine the correct slicing parameters or be able to interpret the results.

The most powerful way to utilize amplitude slice maps is to first view and interpret individual reflection profiles in two dimensions to "get a feel" for the types of reflections present and their depth in the ground. Velocity analysis using hyperbola fitting can be done at this time, if velocities are not already known, so that the depth of visible reflections in the ground can be determined. Slice maps can then be constructed of the number and thickness necessary to produce maps of the reflections that were generated from buried features or stratigraphic layers of interest. Once they are constructed, such as those in plate 2, for many horizontal or subhorizontal slices in the ground, reflection profiles can then be viewed again and reinterpreted to more precisely define the origin of amplitude features seen in the maps. In this way, the orientation, thickness, and relative amplitudes of anomalies are readily interpretable in three dimensions using both slice mapping and two-dimensional profile interpretation.

Amplitude slices are usually made in equal time intervals, with each slice representing an approximate thickness of buried material in the ground. They are always constructed in radar travel times, which can later be converted to depth if velocity analysis has been performed. Viewing amplitude changes in a series of horizontal time slices in the ground is therefore analogous to studying geological and archaeological changes of equal depth layers (Conyers et al. 2002; Goodman et al. 1995; Malagodi et al. 1996; Milligan and Atkin 1993). When amplitude anomalies in each slice are then correlated to known archaeological features and stratigraphic changes that might be available for study in nearby excavations, extremely accurate three-dimensional maps of a site, broken down into levels, can be constructed. In addition, a grid of GPR reflection data may be sliced very thinly and viewed as a video, with slices projected sequentially on a computer screen as layers are "uncovered" from the ground surface to some depth in the ground (Conyers et al. 2002; Grasmueck et al. 2004). The precise location of all reflections in the ground, if processed into many sequential slices, can also be analyzed as a three-dimensional "cube" of data and certain amplitudes rendered to produce realistic images of the subsurface as three-dimensional objects called *isosurfaces* (Conyers et al. 2002; Goodman et al. 1998; Heinz and Aigner 2003; Leckebusch 2003; Leckebusch and Peikert 2001), which will be discussed in more detail later.

Amplitude anomaly maps need not be constructed horizontally or even in equal time intervals. They can vary in thickness and orientation, depending on the archaeological and geological questions being posed. Surface topography and the subsurface orientation of features and stratigraphy of a site may sometimes necessitate the construction of slices that are neither uniform in thickness nor horizontal. They can also be construed to follow one distinct horizon (Conyers and Goodman 1997). This can easily be done on the computer when reflection data are in a digital format.

To compute horizontal amplitude slices the computer programs must compare amplitude variations within reflection traces that were recorded within a defined time window from all profiles in a grid (Conyers and Goodman 1997: 153). Usually computer programs that perform this task will ask for the amount of spatial correlation and interpolation desired for each grid, consisting of many reflection profiles. This process interpolates amplitudes of reflected waves along profiles and between them, producing a grid of data for each slice that can then be mapped. When this is done, both positive and negative amplitudes of reflections are compared to the mean within each slice. Usually no differentiation is made between positive or negative amplitudes, only the magnitude of amplitude deviation from the average. Low-amplitude variations within any one slice denote little subsurface reflection at that level and location and therefore indicate the presence of fairly homogeneous material. High amplitudes indicate the presence of significant subsurface discontinuities, in many cases buried features, as they are produced by reflection at the interfaces between highly contrasting material types. Degrees of amplitude variation in each slice are then assigned arbitrary colors or shades of gray along an ordinal scale, which can be varied to enhance higher, middle range, or lower-amplitude areas at each slice in the ground. Usually there are no specific amplitude units assigned to these color or tonal changes, as "natural" amplitudes are usually preadjusted prior to collection or modified after data are gathered by range gaining or other data processing steps. A crude form of amplitude adjustment, similar to range gaining, can also be performed after the maps are made by altering the range of values that are assigned different colors or shades of gray in each slice map. This will effectively enhance some amplitude values, while suppressing others, to make some features more or less visible to the human eye.

It must be remembered that because most subsurface layers are not perfectly horizontal, and most stratigraphic units vary in thickness laterally, some horizontal time slices may not be comparing reflection amplitudes within the same soil or sediment units in the ground (Beres et al. 1999). For instance, if the programmed amplitude slices cross stratigraphic boundaries of units that are dipping in the ground, there will be an amplitude anomaly registered where the slice crosses the stratigraphic boundary (plate 3). The resulting amplitude map derived from this slicing geometry would therefore be illustrating subsurface changes that are the product of the slicing method and the geologic changes across slices, and not the presence of the types of buried archaeological features that are the target (plate 4). When this occurs, amplitude slice maps can be potentially very misleading and are not illustrating either meaningful geological or archaeological changes. Some interpreters have attempted to characterize the shapes and sizes of anomalies produced by slices that cross-cut geological layers of different sorts, with little success (Beres et al. 1999).

Topographic and cross-cutting complications can often be adjusted for if the orientation and thickness of the subsurface layers are known, which can usually be determined by interpreting processed reflection profiles before slicing. The slicing problem illustrated in plates 3 and 4, where amplitude slices cross stratigraphic boundaries, demonstrates how multiple interpretation techniques including slice mapping and profile interpretation should always be employed in an iterative fashion.

The spacing of profiles and the amount of interpolation between profiles during slice map construction will often determine the resolution of the resulting reflection amplitudes when plotted in map form (Neubauer et al. 2002). If too much interpolation over a large search area is used, amplitude maps will tend to become blurred and feature definition decreases (plate 5). In contrast, a very small search radius and therefore little spatial interpolation can sometimes create a very "noisy" or "busy" amplitude map, and features may remain hidden in the clutter, if the reflections derived from features of interest are not distinct enough from those generated by the surrounding matrix.

A great deal of thought must also go into data processing of individual reflection profiles prior to slicing. Often reflection hyperbolas must be migrated and reflection data filtered before the amplitude slicing step. Then slicing parameters must be chosen including slice thickness, orientation, interpolation radius, and gridding method. Some slicing parameters might generate very accurate and usable maps, while others generate outcomes that may be totally uninterpretable, depending on the data quality, orientation of features in the ground, and stratigraphic complexity. With both experience and experimentation, the infinite number of potential processing parameters can be reduced to a few that have been found to work in certain conditions.

Often many different slicing methods must be attempted and studied before a usable map is obtained. In addition, each grid of GPR data collected during different field projects, each with different equipment and acquisition settings, will always necessitate different types of amplitude analyses using different slicing parameters. This tends to make GPR data collection, processing, and interpretation complicated, but exciting and rewarding, too, as each can potentially yield different results depending on the processing techniques. When this occurs, it is important to determine whether the maps are illustrating what is really present in the ground, or whether one is creating anomalies that are a function only of the slicing techniques used. This can only be done by going back and viewing the individual reflection profiles and visually correlating the reflections visible there to those plotted in the amplitude maps.

Amplitude Slice Maps on Level Ground

When the ground surface and underlying units are horizontal, or nearly so, amplitude time slices that are constructed parallel to the ground surface will usually follow stratigraphic and soil layers and therefore not cut across bed boundaries. It is only when the slices cross bedding planes that anomalous amplitude regions are produced, as illustrated in plates 3 and 4. As long as bedding units are parallel and there are no cross-cutting relationships between units, such as cut and fill channels or intrusive anthropogenic features, each horizontal slice will produce images of archaeological features that are relatively older with depth. Amplitude time slices can then be representative of relative age, as deeper slices will show features that were constructed prior to those visible in shallower slices.

At a historic site in Albany, New York, maps of the city were available showing the location of buildings present on town lots, going back to the year 1857. Amplitude slice maps were then constructed in 25-centimeter depth slices (after velocity analysis was performed), and images of the buried features visible in the GPR maps were compared to the historic lot maps (plate 6). The slice from 50- to 75-centimeter depth shows the building foundations whose location compared almost exactly to domestic structures and a large kiln that were present in 1890. In progressively deeper slices, those 1890 buildings were still visible, but deeper foundations from older structures were also visible in the slice in the 150- to 175-centimeter depth (plate 6). When the locations of those features were compared to the oldest historic maps from 1857, no correlation was found to any mapped structures. The deeper slices were therefore producing images of buildings that were present prior to the construction of any maps of the city. They are awaiting excavation, and their age and function remain unknown.

In this example from historic New York, the horizontal amplitude slice maps can be a way to not only map structure locations over time but, when integrated with enough other information such as historic maps and artifacts from excavations, their function as well. The changing makeup of historic neighborhoods can potentially be determined using these GPR amplitude maps, each of which denotes features from a specific time period, if the sequential amplitude slices are roughly comparable to those building phases. In this way, GPR images can be much more than just a tool for finding and mapping buried features; they can be a database from which to study social and urban change and a wealth of other historic and anthropological questions.

Amplitude Slices on Uneven Ground

When the ground surface changes a good deal over a grid, it is important to collect elevations within that grid so that profiles, and amplitude time slices, can be adjusted for topography. This is especially important when stratigraphic boundaries do not parallel the ground surface but may follow some other orientation. For instance, if a small burial mound was constructed by piling up horizontal soil in layers, amplitude time slices that were produced parallel to the present ground surface would cross many of those bed boundaries, creating anomalous amplitude values that would be meaningless in the resulting maps.

To test this concept, two series of amplitude time slices were constructed at a geophysical archaeology test site in Illinois (Isaacson et al. 1999). At this site, two burial crypts were constructed in 1998 to simulate human burials (even including pig carcasses to simulate human remains). A mound was then constructed of roughly horizontal layers of soil over the crypt. Reflection profiles were collected using 400-megahertz antennas in 25-centimeter spaced transects over the mound. The easiest way to produce amplitude slice maps of this feature is to construct each slice parallel to the ground surface, but by doing so, the slices cross-cut many of the boundaries between horizontal soil layers in the mound (figure 7.9). The resulting 25-centimeter thick amplitude slices constructed in this way produced a series of concentric high amplitude areas in many of the slices, with each high-amplitude ring in the maps denoting the location where those slices crossed the horizontal bedding boundaries (plate 5). Those slices that are not corrected for topography are therefore illustrating amplitudes that have no validity to what is buried below the surface and are the function of incorrect slicing geometry.

When data from the same grid is sliced horizontally after topographic corrections, each amplitude time slice is parallel to the bedding surfaces created by

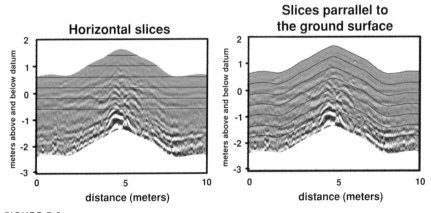

FIGURE 7.9
Slicing Choices in Topographically Complex Areas. This reflection profile over a mound with significant topographic relief can be sliced horizontally or in slices parallel to the ground surface. The slices parallel to the ground surface cross bedding boundaries and produce anomalous reflection readings in slice maps. When profiles are adjusted for topography and then sliced horizontally, each slice is parallel to the horizontal bed boundaries, producing a more realistic series of slice map images, which are shown in plate 7.

the mounding up of soil over the crypts. The slices do not therefore cross horizontal soil units, and no anomalous amplitude readings are created (plate 7). The deeper slices cross the crypts themselves, and high amplitude anomalies are produced in the exact location of the void spaces and crypt edges within the mound.

To create the topographically adjusted slices, the mound had to be topographically surveyed with a transit, with each survey point used to adjust reflection profiles for surface elevation changes. This can be a laborious process, but necessary, if detailed and accurate maps are to be produced. Some GPR systems are in development that will automatically collect surface elevations as well as horizontal locations in a grid using Global Positioning System (GPS) technology, which will allow immediate data corrections of this sort.

Subtle Feature Discovery with Amplitude Mapping

Often reflection profiles can be difficult to interpret, even after filtering, postacquisition processing, and the production of many profile views with differing vertical and horizontal exaggeration. There is often a temptation when one looks at reflection profiles such as the one in figure 7.10 to give up and call the survey a failure as no amplitude changes are readily visible in it. The profile in figure 7.10 was collected with 900-megahertz antennas in a boggy area in the California Sierra Nevada Mountains. The goal was to map recently deposited sedimentary units in the hope of defining fluvial, marsh, and floodplain sediments that might have been present in the mid–nineteenth century. Historic records indicated that the ill-fated Donner party, a wagon train that was immigrating to California and attempting to cross the mountains in November 1846, camped near a creek in the study area and were stranded there all winter. Many eyewitness accounts reported that the survivors found themselves in a bog in the spring of 1847 when the snow melted. Finding the remains of that camp today is complicated because the environment has changed a great deal, and it is now on the edge of a reservoir that was flooded in the 1960s. It was hoped that an analysis of the historic environment as it existed during the time of the Donner party encampment might yield clues to where the winter campsite was located, as it was known there was a small creek nearby in the early winter, and the area became a bog in the spring. The GPR method was considered because even though almost all the prospective area is today wet and boggy, similar environments with an abundance of peat beds had proven excellent areas for GPR mapping in Scotland (Clarke et al. 1999; Leopold and Volkel 2003).

INTERPRETATION OF GPR DATA 157

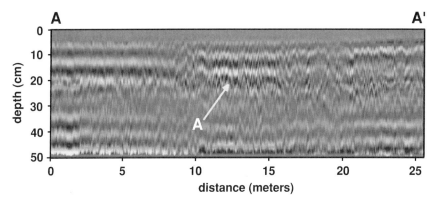

FIGURE 7.10
Reflection Profile with No Distinctive Reflections. This profile was collected in a shallow wetland area near Truckee, California. Little variation is apparent in the reflections, but some changes in amplitude are noticeable as an area of small reflection hyperbolas (A). This profile (A-A') is located on the amplitude slice map in plate 8.

The reflections in the 900-megahertz profiles that crossed the present-day bog proved to be noisy and discontinuous, and few good reflections were visible that could be readily interpreted (figure 7.10). There are, however, changes in the reflection character along profiles, with some areas containing few good reflections and others that appear to contain many very small hyperbolic reflections, especially within the upper 20 centimeters of the reflection profiles. As little interpretation could be done using the individual profiles, it was decided to study the amplitude changes spatially within all the profiles in the grid to determine if there were any patterns to the distribution of these reflections either spatially or with depth.

When the amplitudes in all the profiles in the grid were studied in slice maps, one sinuous area of higher-amplitude reflections was visible in the 10- to 20-centimeter slice (plate 8). After further study of the reflection profiles in two dimensions, it was hypothesized that the anomalously high-amplitude area corresponded to the presence of many small gravel clasts in a creek, each of which generated the small reflection hyperbolas at that depth. Auger holes were then dug on either side of the high-amplitude feature and within it (plate 8). A study of the sediments recovered from them showed that holes 1 and 3 consisted of silt and peat, with abundant charcoal, while the sediment in hole 2 was mostly sand and gravel and contained very little peat. This subsurface information confirms that the sinuous anomaly in the 10- to 20-centimeter slice represents a small sand- and gravel-filled creek channel. The areas adjacent to it, which are much lower in reflection amplitude, are areas where marsh and floodplain sediments were deposited.

Although remains of the Donner party camp were not found in this immediate area, the study was successful in defining the shallow creek with adjacent marshy floodplain deposits, which can be used as a guide for further subsurface testing in a search for artifacts. Most important, this study illustrates how even reflection data that are difficult to interpret in two-dimensional profiles can produce useful data when studied in amplitude slices. When data of this type are then incorporated with standard archaeological and historical information, large areas of ground can be studied quickly, and excavation efforts can be concentrated in the most prospective locations.

Often GPR amplitude slice maps are capable of producing images that are not only almost invisible in reflection profiles, as shown in the prior example, but the buried features that produced the reflection anomalies are also almost invisible to the human eye even when uncovered in excavations. A GPR study of this sort was conducted in an orchard, where surface plowing had destroyed any indication of buried features likely to exist below. The area surveyed was the site of an early homestead in the mid-1800s in Denver, Colorado, which was converted to a stage wagon stop and finally reverted to a family farm in the twentieth century. Historic documents indicate that a number of buildings had been located somewhere in the orchard area, but their exact locations were unknown. The area had also been subjected to a number of historic floods, which buried any possible remaining architectural features below more than a meter of sediment.

A grid of 400-megahertz GPR reflection data was collected in the orchard, and horizontal amplitude slice maps were constructed every 25 centimeters in the ground, after radar travel times were converted to depth (plate 10). At the 75- to 100-centimeter depth, a distinct linear feature was discovered, and in the deeper slices, another linear feature crossing it at an angle. Modern utility maps show plastic water lines cutting through the orchard, which generated those linear reflection anomalies. More interesting, however, was a 4-meter square high amplitude feature in the 75- to 100-centimeter depth, which was not correlative to any of the historic buildings that had been mapped in the vicinity. This feature was hypothesized to be a buried building floor, if for no other reason, than its perfectly square geometry.

Auger holes were dug both inside and outside the square feature, and no discernible difference could be seen in the two sediment and soil samples from the depth indicated in the amplitude slice map. Thinking that perhaps velocities, which had been estimated from hyperbola fitting of point source hyperbolas generated from the pipes, were incorrect, researchers placed a test excavation directly on top of one of the pipes to uncover it in order to confirm depth to a known object. The pipe was found at exactly the depth shown in the GPR maps, raising the confidence level of the velocity analyses and the resulting amplitude slice map depths. A more careful analysis of the soil and sediment stratigraphy in the excavations adjacent to and just below the pipe was then made. A 2-centimeter thick layer that was just a little sandier than layers above and below was found within the square GPR feature. When excavations were extended outward, this sandy layer was found to be overlain by broken pieces of flat sandstone, which were probably used as pavers in the floor of a small house or shed.

It appears that there was a small building in the orchard at one time, whose floor was paved with sand covered with flat stones. When the building was abandoned, the usable stone pieces were probably salvaged for use elsewhere, and the remaining sandy subfloor was covered by sediment during floods and by the buildup of soil. All that remains of the house today is the very subtle sand layer. The feature is so subtle that normal excavation methods would have likely overlooked it, as it would have been interpreted as just another sandy layer in the orchard's sediment and soil package. Even once the feature had been discovered in excavations, it could still not be seen in GPR reflection profiles without using a good deal of imagination. Only subtle changes in radar reflection amplitudes, processed as amplitude slice maps, were capable of finding and mapping this feature, which was only distinguishable by its distinctive square shape.

Amplitude Maps to Search for Vertical Features and Graves

Most often GPR mapping in archaeology is used to produce images of planar features such as buried house floors, important stratigraphic interfaces that are subhorizontal, or distinct geometric orientations of point source reflections, such as buried wall tops or stone circles. Vertical interfaces that might be found in grave shafts or other similar features are often more difficult to map, as most radar energy propagating into the ground from surface antennas is traveling parallel to the boundaries of interest, and little reflection therefore results. Vertical shafts that lead to tombs, however, are visible when amplitudes are mapped in slices, as noticeable reflection changes between the material that fills the shafts and that surrounding it. In Japan, shafts leading to burial tombs were visible as high-amplitude reflections in numerous slices stacked on top of each other

(Conyers and Goodman 1997: 184). Historic and prehistoric graves are also often visible in the same fashion (Bevan 1991; Davenport 2001a, 2001b; Davis et al. 2000; Nobes 1999; Strongman 1992).

In a historic cemetery in Colorado, many graves could be located by amplitude changes in a number of horizontal slices (plate 9), although these changes were difficult to identify in individual profiles (figure 7.11). In this case, the graves were found by analyzing the lowest amplitudes, as the grave shafts were filled with homogenized back fill material. This occurred because during their excavation, the natural stratigraphy in the ground was destroyed, producing fewer high-amplitude reflections in the refilled shafts than the surrounding intact units.

If graves are excavated into sediment that is layered, the layers outside the grave shaft will retain their natural stratigraphy, while the material placed back into the grave will be homogenized, producing a distinctive disruption in layering (figure 7.11). Sometimes these truncation features are visible in amplitude slice maps, but often profiles must be processed and analyzed individually, and the location of these features hand-plotted on maps as they are identified in each reflection profile. Any human remains preserved within them will probably not be visible, as they may not contrast enough with the surrounding material.

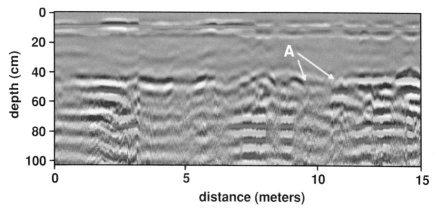

FIGURE 7.11
Stratigraphic Truncation in a Cemetery. This reflection profile was collected in an historic cemetery near Boulder, Colorado, where only the vertical shaft is visible as an area of low amplitude. Truncations of the stratigraphy by the vertical shaft are visible at locations (A).

When graves contain coffins that have not collapsed and retain some void space, they are readily visible in profile as distinctive reflection hyperbolas (figure 7.12). The hyperbolic reflections are often high in amplitude, and if the hyperbola axes are migrated back to their sources, very distinct images of graveyards can be made, with various coffin sizes and depths of burials differentiated. Often reflection profiles and amplitude slice maps in these cemeteries are distinct enough to determine the difference between adult and child burials by the size of the coffins and depth of burial. Whether coffins were lined with metal can also be determined by whether they have the distinctive multiple reflections common for buried metal.

Production of Rendered Images

The unique ability of GPR systems to collect reflection data in a three-dimensional package lends itself to the production of a number of other three-dimensional images not possible using other methods (Conyers et al. 2002; Goodman et al 1998, 2004; Heinz and Aigner 2003; Leckebusch 2000, 2003). If reflection data are collected in a grid of closely spaced transects, and if there are many reflection traces gathered along each transect, reflection amplitudes can be accurately placed in three dimensions and then rendered using a number of visual display programs. In this way, GPR data from archaeological sites become analogous to many other imaging techniques used in other disciplines, which

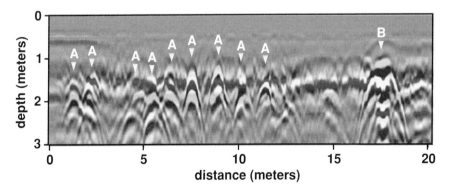

FIGURE 7.12
Graves with Distinct Reflection Hyperbolas. Each casket (A) is visible as a distinct reflection hyperbola, probably generated from void spaces within. A very large casket, composed of or lined with metal, is visible at location B. Reflection data were collected at the Military Cemetery at Fort Vancouver, Washington.

rely on energy sources such as sonic waves and magnetic resonance. In medical imaging complex, three-dimensional techniques can produce images of certain amplitudes derived from these energy sources to display internal body parts, or even electrical impulses in the brain as a function of different stimuli. In archaeology, radar reflections can be used in the same way but instead produce images of buried cultural features.

Using GPR data, buried features or interfaces can be rendered into isosurfaces, meaning that the interfaces producing the reflections are placed in three dimensions, and a pattern or color is assigned to specific amplitudes in order for them to be visible (Heinz and Aigner 2003; Leckebusch 2003). In programs that produce these types of images, certain amplitudes (usually the highest ones) can be patterned or colored while others are made transparent. Computer generated light sources, to simulate rays of the sun, can then be used to shade and shadow the rendered features in order to enhance them, and the features can be rotated and shaded until a desired product is produced.

These types of images have been denigrated by some as "too flashy," "without any practical value," or more like high-tech video games than archaeological geophysics. Humans, however, are visual animals, and we can often comprehend three-dimensional images much easier than reflections in profiles or standard amplitude slice maps. One of the goals of archaeological geophysics should be to "see into the ground," and what better way than to make an image of the ground that is representative of what a site would look like if totally excavated? Three-dimensional renderings can often be the most readily comprehensible of all GPR images for this purpose, especially for the nongeophysically initiated.

To produce rendered images with GPR profiles, all reflection profiles in a grid must first be processed and filtered to produce the "cleanest" final product possible. Background noise and interfering frequencies must be removed and important amplitudes from features to be rendered range gained to enhance their visibility. Hyperbolic reflections should also be migrated back to their origins. Many slice maps must then be produced in very thin time intervals, in order to produce layers of digital data in closely spaced parallel planes. If these slices are constructed too thinly, reflection waveforms can sometimes be dissected into many small meaningless packages, and the resulting amplitude values may contain only a part of each wave reflected, and not the wave as a whole. For instance, if there is a 20-nanosecond window over which reflection data were collected, and each reflected wave in the ground has a wavelength of about 2 nanoseconds (about average for a 400-megahertz antenna), then the most slices that should be

INTERPRETATION OF GPR DATA   163

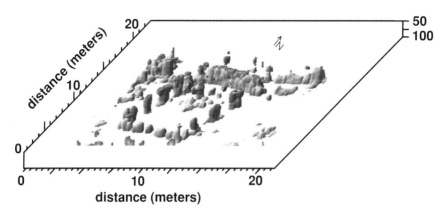

FIGURE 7.13
Three-Dimensional Rendered Surface. This is a three-dimensional rendering of the buried building at Petra, Jordan, shown in the slice maps in plate 2. The highest-amplitude reflections are rendered in their three-dimensional location, and artificial light is projected to make the features more visible.

generated (and not dissect the waves into many pieces to obtain amplitudes) would be 10. For this reason, a running average may be preferable for creating a large three-dimensional database for rendering, where a 20-nanosecond window of reflections is dissected into 40 horizontal slices, each 3 nanoseconds in thickness, but overlapping the adjoining slices by a nanosecond or two. There would be some averaging of the amplitude data in this method, but a greater likelihood that complete waveforms will be averaged in each slice, and then made visible by rendering. Grasmueck et al. (2004), however, have shown that in some contexts very thin slices with no overlap can still produce good images.

When the buried structure from Petra in plate 2 was sliced thinly in this fashion, and only the highest amplitudes were rendered into an isosurface, a very distinct image of a buried building was constructed (figure 7.13). In this rendering, the shallowest slices were not included, as they contained many small point source reflections from shallow rock rubble. The very deepest slices were also not included, as the data from that depth tended to be noisy and resulted in high-amplitude streaking in the rendered images (Conyers et al. 2002). These reflection data were ideal for rendering because most of the reflections collected in the grid were generated from buried architecture, and the surrounding wind-blown sand matrix generated almost no reflections of any amplitude. Once the rendering was constructed on the computer, it could also be rotated or tilted in a number of different orientations in a video display (Conyers et al. 2002).

8
Conclusion

Ground-penetrating radar can be one of the most complicated of near-surface archaeological geophysical techniques, but also one of the more rewarding, as it has the ability to map what is buried in the ground in three dimensions. Conditions, although, must be conducive for radar energy propagation in the ground, features must be distinct enough to be differentiated from the background sediment or soil, and at a depth that can be resolved with the equipment available. For all these variables to be met, the geological and archaeological features at each site must be physically understood so that the origins of the reflections in the ground can be correctly interpreted. In addition, an understanding of how reflections are created, and how acquisition and processing steps can enhance, or sometimes obscure, these reflections is crucial.

Huge advances in GPR acquisition and processing hardware and software have been made in the last few years, which can transform what used to be considered unusable or at best marginal reflection data into important maps and images of the subsurface. One of the most important of these advances is amplitude analysis, which can process tens or even hundreds of reflection profiles at once, creating usable databases from massive amounts of reflections within hours, or even minutes, of completing a survey. Without this type of data analysis, GPR data must be interpreted using visual analysis and hand mapping of reflections in many profiles, which can prove daunting for all but the most motivated researcher with a good deal of free time. But even though computer-processing techniques are faster and more efficient, the primary database (each reflection trace in each profile) must have been collected and processed correctly before undergoing this powerful data transformation step. The fun and reward always comes in the data processing, but unless good-quality reflection data are originally collected, and a great deal of care is taken in postacquisition processing, one can never be assured of high-quality (or at least usable) results.

In GPR data collection, a great deal of time must be devoted to choosing correct antennas for optimum depth resolution, setting up grids to make sure target areas are completely covered, and choosing correct setup parameters to optimize recorded data quality. A proper calibration of equipment for the infinite number of variables that can occur within buried geological and archaeological materials must be thoughtfully applied. Each site will present its own set of both equipment problems that will have to be overcome, as well as soil and sediment variations that need to be adjusted for, if final results of GPR mapping are to be at all useful. Most important, an understanding of "why" data were collected in a certain fashion is necessary, if profiles and maps are to be interpreted correctly. It is not at all useful to be able to say that a set of acquisition procedures and setup parameters were used because the data "looked good," as this observation does not lend itself to being able to say "why" the final products appear as they do. To make results explainable for the more general archaeological audience, an understanding for each site of how GPR energy is generated, transmitted, reflected, attenuated, and then finally recorded is crucial. Then each acquisition and processing step, and the final mapping and imaging procedures, must be also explainable in an understandable way, or the results of this powerful but sometimes complicated near-surface geophysical technique will always be in doubt.

Although most GPR practitioners usually "show off" only their prettiest amplitude slice maps and reflection profiles in the published literature or when giving talks, while ignoring the harder-to-interpret maps and profiles, all the data acquired should be considered a primary database, and they must be correctly understood for the final products to be correctly interpreted. A combination of profile interpretation, selective data processing, amplitude slice–mapping, and reinterpretation of reflection profiles comparing them to the slice maps is always necessary. Often amplitude slice maps are produced first so that interesting distributions of reflections of certain amplitudes in the ground can be immediately seen. But the origin of those amplitudes must be understood by studying each of the reflection profiles and relating the origin of reflections to what generated them in the ground, or mistakes in interpretation will likely result. So, even though GPR data today can be processed into slice maps many orders of magnitude faster than just a few years ago, the "old-fashioned" methods of profile reflection analysis must still be a part of the final interpretation process.

When poor reflection data are acquired, it is crucial to try to understand what happened and then learn from those failures. With GPR, the number of variables, some of which can be controlled for, and others that cannot, are infinite. If archaeological geophysicists are to overcome the stigma of being "operators of black boxes" that produce strange wiggles, which only they and others in the "geophysics club" can understand, then we must all try to understand our data, no matter what its quality, and be able to explain to others both our successes and failures.

Many years ago, a colleague working at a deeply buried Paleoindian site in Texas hired some geophysicists to conduct a GPR survey to define important buried stratigraphic layers. The consulting geophysicists spent all day working with their equipment in an intense but quiet fashion, collecting reflection data, and presumably interpreting the results in the back of their van. At the end of the day, they proclaimed to the archaeologists, who were expectantly waiting for results, that "GPR doesn't work here." They quickly loaded up their equipment and drove off in a cloud of dust, never to be seen again. The archaeologists were struck dumb, as they knew little about the GPR method, and were left with no results and only a fairly large invoice for services rendered. To this day they remain unenlightened about what, if anything, was accomplished that day, and, needless to say, they have never again used GPR. It is negative experiences such as this that we must do our best to keep from happening. The only way for this to occur is by first understanding all acquisition, processing, and interpretation methods, and then correctly and patiently explaining their outcomes to others.

It is also important to emphasize that archaeologists cannot arbitrarily employ the GPR techniques presented in the book without having to break the ground with a shovel and get on their hands and knees with a trowel and dustpan. Ground-penetrating radar analysis will never be able to replace standard archaeological methods and to be most successful the method should be integrated with them. Subsurface radar reflections will never be able to determine the age of an archaeological feature, what kind of pottery it may have in context with it, or the color of the pigment it is decorated with. Only excavations can yield this type of information. The method's greatest strength lies in its ability to discover and create accurate images of hidden features in three dimensions and produce maps and profiles of important stratigraphy between and surrounding standard archaeological excavations.

Many in the archaeological community may continue to employ GPR only as an "anomaly-finding" device to locate possible features that can later be excavated. Although this may remain an important usage, in the future GPR's maximum effectiveness will be when it can be integrated with detailed archaeological and geological information collected from excavations and stratigraphic studies. When this is done, the method becomes a valuable tool to accurately map the anthropogenic and natural environment of a site and to integrate those maps with other more standard archaeological data. Without knowledgeable integration of GPR products with information from archaeologists and stratigraphers working at any one site, GPR will remain only a "black box" method that creates interesting (and possibly important) two- and three-dimensional images of features under the ground.

The GPR method is also only effective as a mapping tool in certain environments under specific soil and moisture regimes. Its success or failure is premised on the knowledgeable application of the correct equipment, the appropriate acquisition parameters, and the interpretative ability of the archaeologist, geologist, or geophysicist. It is one matter to recognize GPR anomalies (which may or may not have significance) and quite another to correctly interpret the most important reflections in order to derive archaeological or environmental meaning.

In an attempt to specify some of the successes and failures of GPR, an assessment of the feasibility of imaging various features in the ground is presented in table 8.1. These qualitative assessments are derived from the published literature and personal experience, and it is hoped that others using slightly different techniques may produce better results for some of what might be considered the more marginal archaeological applications. In addition, this book has focused on only a very small fraction of the sites and conditions around the world where the method could be potentially applied. Also, each of the databases used to produce interpretations in this book were collected during specific times and under conditions that can never be perfectly replicated.

Usable GPR data have recently been collected in some areas or conditions where "GPR dogma" has in the past declared that no usable results could possibly be obtained, showing that the method has much broader applicability than previously thought. Some of those results are illustrated in this book. Table 8.1 must therefore be used only as a general guide for accessing the feasibility of GPR and can in no way be relied on as an exclusive guide to all situations.

The general feasibilities described in table 8.1 must be tempered by an understanding that the depth of targets as well as local ground conditions can drastically alter radar energy penetration depth and reflection amplitudes. A common misconception is that resolution of buried archaeological features or important stratigraphic layers is better near the surface and decreases with depth. Generally this is true, but in some cases, if the features of interest are located within the antenna's near-field zone, they may remain hidden. Processing steps can sometimes be applied to those obscure but important shallow reflections, and they can possibly be coaxed out of the database. These same features, however, if buried deeper in the ground might be visible with the same transect spacing due to the conical spreading of the radar beam that illuminates an ever-increasing surface area with depth. Usually the closer the spacing of transects within a grid, the better definition of buried materials, irrespective of antenna frequency.

Many believe that GPR surveys can only be conducted on level or nearly level ground. While it is always easier to work on a clear flat surface, uneven or rough ground should not preclude using GPR. Some successful surveys, conducted in areas with severe topographic variations, often yield surprising useful results. If detailed surface elevation measurements are obtained over the grid, profiles can be corrected for topography and archaeological features that would otherwise be difficult to discern in standard profiles may become apparent. On rough ground, if the recorded reflection traces are stacked in order to average out the minor surface disturbances, and background removal filters are applied to the data after collection, good reflection data can still be obtained. Lower-frequency antennas (less than 300 megahertz) are much less influenced by small changes in their roll and pitch when pulled over rough surfaces, which is not the case when using higher frequency antennas that are smaller in size. Even small surface irregularities that are not factored out by topographic corrections can severely disrupt subsurface reflections obtained from high-frequency antennas. No matter what the antenna size or frequency it is imperative to keep antennas in the same general orientation with the ground, or energy coupling changes will cause changes in the nature of reflected waves that can be confused with "real" changes in the ground.

It is extremely important that field acquisition time be allotted for equipment calibration and velocity tests prior to conducting a survey. When arriving at a site, it is always tempting to immediately begin acquiring reflection data before the field conditions are fully understood. This is especially true if the GPR equipment is being rented by the day or hour. Haste of this sort can many times yield unfortunate results, especially if it is discovered after returning from the field that important equipment adjustments such as trace stacking or time window adjustments were incorrectly made. The same is true with time–depth conversions that can only be

Table 8.1. Feasibility of Using GPR to Discover and Map Some Buried Archaeological Features and Stratigraphy

Archaeological Target	Feasibility for GPR	Reasons for Assessment
Pit dwelling filled with different material than the surrounding matrix	Good	Good velocity contrast between floor and matrix will produce strong reflections.
Trenches, buried moats, hollow tunnels	Excellent	Good velocity contrasts at interfaces with surrounding material of the void.
Buried excavation trenches	Moderate to good	Good velocity contrast when backfill materials are different than surrounding materials or are less compacted.
Fire pits with baked bottoms greater than about 1–2 meters in diameter	Good	Firing can create a baked surface that readily reflects radar energy.
Fire pits less than one meter in diameter	Moderate to poor	Usually too small to be visible unless pits are very shallow and high-frequency antennas are used.
Stone foundations buried in fine-grained material	Excellent	Vertical walls will create reflection hyperbolas from their tops and sides.
Clay, stone, or wooden structures buried by rocky material	Poor	Too many small reflections (clutter) are produced from the rocky material and can be confused for the features of interest.
Clay, stone, or wooden structures buried by wet clay	Usually poor	The high electrical conductivity of some clay will severely attenuate radar energy.
Clay, stone, or wooden structures buried by moist or dry fine-grained volcanic material	Good	Good velocity contrast between the structures and the surrounding material creates strong reflections.
Kiln floors or roofs	Excellent	The high temperatures in the kilns baked the surrounding material, creating an excellent radar reflection surface.
Buried living surfaces overlain by material of a different lithology	Good	An aerially extensive interface with a good velocity contrast will readily reflect radar waves.
Small stone tools dispersed in soils	Poor	Often target objects are too small to reflect enough radar energy to be visible. This may be overcome if a high enough frequency antenna is used.
Small metallic tools dispersed in soils	Moderate	If objects are not buried too deeply and high-frequency antennas are used, metal objects will create small but visible reflection hyperbolas; can also create very visible multiple reflections in profiles.

Feature	Rating	Description
Moderate to large metal objects	Excellent	Metal is a perfect radar reflector and will generate visible reflection hyperbolas, as long as they are not too deeply buried.
Small clay artifacts	Poor	Usually do not create a large enough velocity contrast with surrounding material to be visible as a reflection
Aerially extensive concentrated areas of pottery sherds	Moderate	Depending on their thickness and velocity contrast with the surrounding matrix, they can create a noticeable reflection that appears as one layer.
Burials filled with material that is different than the surrounding matrix	Moderate	Good velocity contrast at the interface of the two different materials may create good reflections; visible if burial is large and not too deeply buried.
Stone- or clay-lined burial crypts in fine-grained matrix	Good	Rock and clay usually make a good contrast with the surrounding material and burials are usually large enough to be visible with most radar frequencies.
Features within rock-lined chambers	Poor	Cap stones and rock lining will reflect most of the radar energy before it can enter the chamber.
Compacted mud or soil walls and floors buried by fine-grained material	Good	Good velocity contrast exists at the wall interfaces will create reflections.
Stratigraphic layers with thicknesses less than the transmitted radar wavelength	Poor	The top and bottom of the layer cannot be resolved because the wavelength of the transmitted energy is too long.
Stratigraphic layers thicker than the transmitted radar wavelength	Good	Both the top and bottom of the layer will reflect energy from the same transmitted waves.
Any feature below a thick, wet clay layer	Poor	Many thick wet clay units are very electrically conductive and will attenuate most, if not all, radar energy; but there are some notable exceptions, as not all clay is conductive.

accurately made if velocity measurements are obtained as part of a field data acquisition program at the same time as the reflection profiles are collected.

Each GPR survey will produce unique results, depending on the field conditions at the time the survey was made. Often there are variations in data quality in the same area from day to day. Soil and sediment moisture and other environmental factors can vary with moisture changes and surface conditions, and radar wave velocities and subsurface resolution will vary accordingly. If velocity tests are not made while in the field conducting the survey, they can never be accurately replicated, and after the fact, time–depth corrections can sometimes be only a matter of conjecture.

An understanding of the subsurface stratigraphy as it relates to the generation of reflections is also crucial in the interpretation process. This can only be accomplished if the stratigraphy in test excavations or outcrops is visible and studied. If possible, some GPR profiles in a grid should be acquired that can be "tied" to visible stratigraphy so that the reflections obtained in the profiles can be correlated directly to buried horizons of interest. If this is not done, the origin of individual reflections will always be in doubt, and some possibly important features may go unrecognized. In some situations, it may be difficult to get permission to excavate in order to observe and measure the stratigraphy, especially if the purpose of the survey is to map a site noninvasively. If excavation is not allowed in an archaeologically sensitive area, it may still be possible to extend survey transects to nearby road cuts or to excavations that are located away from the site but in a similar geological setting. As a last resort, it may be permissible to obtain stratigraphic information from auger probes or small diameter cores that may be correlated to reflections in GPR profiles.

Users of GPR should always take into account information about soil and sediment conditions from published reports or consult with geologists or soil scientists who have a familiarity with the study area. If soils are very clay-rich, especially when wet, or the features of interest are located quite deep in the ground, other geophysical techniques may be more useful than GPR. Recent work, however, has shown that even wet clay, if it is not electrically conductive, will still allow the passage and reflection of radar energy to surprising depths (Conyers 2004). Although many other geophysical mapping methods can still yield meaningful data where GPR fails, GPR is one of the only near-surface methods that can be accurately calibrated to give reliable depth information and produce three-dimensional maps.

CONCLUSION

During data acquisition, it is important, but many times difficult, to refrain from making judgments about the success or failure of the survey. It is always tempting to expound to onlookers about the origin of reflection anomalies that may be visible as raw profiles on the computer screen in "real time." Thoughtful processing and interpretation of the data after returning from the field is usually necessary for any accurate assessment of a survey's success. Preliminary conclusions based on a perusal of raw reflection profiles in the field are often inaccurate and any evaluation of the survey's success based on them can be both hasty and potentially embarrassing. After applying background removal filters, amplitude slice analyses, and other interpretative techniques to the reflection data, otherwise invisible features often appear in the most unlikely places. A detailed analysis of the data is almost always necessary and casual approaches to data processing and interpretation will likely result in flawed or failed surveys.

For stratigraphically or archaeologically complicated sites, an iterative process of computer manipulation and mapping combined with manual interpretation of profiles is almost always necessary. Individualized and sometimes sophisticated techniques must often be improvised to deal with the complexities of some sites. These processes may necessitate a cooperative analysis by a team of archaeologists, geologists, and geophysicists. This type of approach is often difficult due to economic and time constraints, necessitating that the archaeologist in charge of the survey be well versed in many of the techniques discussed in this book.

References

Aitkin, M. J.
  1958 Magnetic prospecting: 1—The Water Newton Survey. *Archaeometry* 1 (1): 24–29.
  1974 *Physics and Archaeology.* 2nd ed. Clarendon, Oxford.

Annan, A. P., and L. T. Chua
  1992 Ground penetrating radar performance predictions. In *Ground Penetrating Radar*, J. A. Pilon (editor). Geological Survey of Canada, Paper 90-4: 5–13.

Annan, A. P., and S. W. Cosway
  1992 *Simplified GPR beam model for survey design.* Extended abstract of 62nd Annual International Meeting of the Society of Exploration Geophysicists, New Orleans, October 25–29, 1992.
  1994 GPR frequency selection. In *Proceedings of the Fifth International Conference on Ground Penetrating Radar:* 747–760. Waterloo Centre for Groundwater Research, Waterloo, Canada.

Annan, A. P., and J. L. Davis
  1977 Impulse radar applied to ice thickness measurements and freshwater bathymetry. *Geological Survey of Canada, Report of Activities* paper 77-1B: 117–124.
  1992 Design and development of a digital ground penetrating radar system. In *Ground Penetrating Radar*, J. A. Pilon (editor). Geological Survey of Canada, Paper 90-4: 49–55.

Annan, A. P., W. M. Waller, D. W. Strangway, J. R. Rossiter, J. D. Redman, and R. D. Watts
  1975 The electromagnetic response of a low-loss, 2-layer, dielectric earth for horizontal electric dipole excitation. *Geophysics* 40 (2): 285–298.

Arcone, Steven A.
  1995 Numerical studies of the radiation patterns of resistivity loaded dipoles. *Journal of Applied Geophysics* 33: 39–52.

ASTM International
  2003 *Standard Guide for Using the Surface Ground Penetrating Radar Method for Subsurface Investigation.* D6432-99. http://www.astm.org.

Baker, P. L.
  1991 Response of ground-penetrating radar to bounding surfaces and lithofacies variations in sand barrier sequences. *Exploration Geophysics* 22: 19–22.

Balanis, Constantine A.
  1989 *Advanced Engineering Electromagnetics.* Wiley, New York.

Basson, U., Y. Enzel, R. Amit, and Z. Ben-Avraham
  1994 Detecting and mapping recent faults with a ground-penetrating radar in the alluvial fans of the Arava Valley, Israel. In *Proceedings of the Fifth International Conference on Ground Penetrating Radar:* 777–788, Waterloo Centre for Groundwater Research, Waterloo, Canada.

Batey, Richard A.
  1987 Subsurface interface radar at Sepphoris, Israel. *Journal of Field Archaeology* 14 (1): 1–8.

Beres, L., and H. Haeni
  1991 Application of ground-penetrating radar methods in hydrogeologic studies. *Groundwater* 29 (3): 375–386.

Beres, Milan, Peter Huggenberger, Alan G. Green, and Heinrich Horstmeyer
  1999 Using two and three-dimensional georadar methods to characterize glaciofluvial architecture. *Sedimentary Geology* 129: 1–24.

Bernabini, M., E. Brizzolari, L. Orlando, and G. Santellani
  1994 Application of ground penetrating radar on coliseum pillars. In *Proceedings of the Fifth International Conference on Ground Penetrating Radar:* 547–558. Waterloo Centre for Groundwater Research, Waterloo, Canada.

Bevan, B. W.
  1977 *Ground-penetrating radar at Valley Forge.* Geophysical Survey Systems, North Salem, N.H.
  1991 The search for graves. *Geophysics* 56 (9): 1310–1319.

1998 *Geophysical Exploration for Archaeology: An Introduction to Geophysical Exploration.* Midwest Archaeological Center Special Report No. 1. National Park Service, Lincoln, Nebr.

2000 An early geophysical survey at Williamsburg, USA. *Archaeological Prospection* 7: 51–58.

Bevan, Bruce, and Jeffrey Kenyon
   1975 Ground-penetrating radar for historical archaeology. *MASCA Newsletter* 11 (2): 2–7.

Birkeland, Peter
   1999 *Soils and Geomorphology.* 3rd ed. Oxford University Press, New York.

Bjelm, L.
   1980 Geologic interpretation of SIR data from a peat deposit in northern Sweden. Unpublished manuscript, Lund Institute of Technology, Department of Engineering Geology, Lund, Sweden.

Bogorodsky, V. V., C. R. Bentley, and P. E. Gudmandsen
   1985 *Radioglaciology.* Reidel, Dordrecht.

Bridge, J. S., J. Alexander, R. E. L. Collier, R. L. Gawthorpe, and J. Jarvis
   1995 Ground-penetrating radar and coring used to study the large-scale structure of point-bar deposits in three-dimensions. *Sedimentology* 42: 839–852.

Bristow, C. S., and H. M. Jol
   2003 *Ground Penetrating Radar in Sediments.* Geological Society Special Publication No. 211, The Geological Society, London

Bristow, C., J. Pugh, and T. Goodall
   1996 Internal structure of aeolian dunes in Abu Davi determined using ground-penetrating radar. *Sedimentology* 43: 995–1003.

Bruschini, Claudio, Bertrand Gros, Frrederic Buere, Pierre-Yves Piece, and Oliver Carmona
   1998 Ground-penetrating radar and imaging metal detector for anti-personnel mine detection. *Journal of Applied Geophysics* 40: 59–71.

Bucker, Frank, Manuela, Gurtner, Heinrich Hortsmeyer, Alan G. Green, and Peter Huggenberger
   1996 Three-dimensional mapping of glaciofluvial and deltaic sediments in central Switzerland using ground penetrating radar. In *Proceedings of the 6th International Conference on Ground Penetrating Radar:* 45–50. Department of Geoscience and Technology, Tohoku University, Sendai, Japan.

Buderi, R.
1996 *The Invention That Changed the World.* Simon & Schuster, New York.

Butler, D. K., J. E. Simms, and D. S. Cook
1994 Archaeological geophysics investigation of the Wright Brothers 1910 hanger site. *Geoarchaeology: An International Journal* 9 (6): 437–466.

Cai, Jun, and George A. McMechan
1994 Ray-based synthesis of bistatic ground penetrating radar profiles. In *Proceedings of the Fifth International Conference on Ground Penetrating Radar,* 19–29. Waterloo Centre for Groundwater Research, Waterloo, Canada.

Carcione, José M.
1996 Radiation patterns for 2-D GPR forward modeling. *Geophysics* 63 (2): 424–430.

Carrozzo, M. T., G. Leucci, S. Negri, and L. Nuzzo.
2003 GPR Survey to understand the stratigraphy of the Roman Ships Archaeological Site (Pisa, Italy). *Archaeological Prospection* 10: 57–72.

Cassidy, N. J., A. J. Russell, J. K. Pringle, and J. L. Carrivick
2004 GPR-derived architecture of large-scale Icelandic Jokulhlaup deposits, northeast Iceland. In *Proceedings of the Tenth International Conference on Ground Penetrating Radar: June 21–24, Delft, The Netherlands,* Evert Slob, Alex Yarovoy and Jan Rhebergen (editors). Delft University of Technology, The Netherlands and the Institute of Electrical and Electronics Engineers, Inc., Piscataway, New Jersey: 581–584.

Chignell, Richard J.
2004 The radio licensing of GPR systems in Europe. In *Proceedings of the Tenth International Conference on Ground Penetrating Radar: June 21–24, Delft, The Netherlands,* Evert Slob, Alex Yarovoy and Jan Rhebergen (editors). Delft University of Technology, The Netherlands and the Institute of Electrical and Electronics Engineers, Inc., Piscataway, New Jersey: 3–6.

Clark, Anthony
1990 *Seeing Beneath the Soil.* Batsford, London.

Clarke, M. Ciara, Erica Utsi, and Vincent Utsi
1999 Ground-penetrating radar investigations at North Ballachulish Moss, Scotland. *Archaeological Prospection* 6: 107–121.

Collins, Mary E.
1992 Soil taxonomy: A useful guide for the application of ground penetrating radar. In *Fourth International Conference on Ground-Penetrating Radar: June 8–13,*

Rovaniemi, Finland, Pauli Hanninen and Sini Autio (editors). Geological Survey of Finland Special Paper 16: 125–132.

Conyers, Lawrence B.
1995 The use of ground-penetrating radar to map the buried structures and landscape of the Ceren site, El Salvador. *Geoarchaeology: An International Journal* 10 (4): 275–299.

2004 Moisture and soil differences as related to the spatial accuracy of GPR amplitude maps at two archaeological test sites. In *Proceedings of the Tenth International Conference on Ground Penetrating Radar: June 21–24, Delft, The Netherlands*, Evert Slob, Alex Yarovoy and Jan Rhebergen (editors). Delft University of Technology, The Netherlands and the Institute of Electrical and Electronics Engineers, Inc., Piscataway, New Jersey: 435–438.

Conyers, Lawrence B., and Catherine M. Cameron
1998 Finding buried archaeological features in the American Southwest: New ground-penetrating radar techniques and three-dimensional computer mapping. *Journal of Field Archaeology* 25 (4): 417–430.

Conyers, Lawrence B., and Dean Goodman
1997 *Ground-penetrating Radar: An Introduction for Archaeologists*. AltaMira, Walnut Creek, Calif.

Conyers, Lawrence B. and Jeffrey E. Lucius
1996 Velocity analysis in archaeological ground-penetrating radar studies. *Archaeological Prospection* 3: 312–333.

Conyers, Lawrence B., and Hartmut Spetzler
2002 Geophysical Exploration at Ceren. In *Before the Volcano Erupted*, Payson Sheets (editor). University of Texas Press, Austin: 24–32.

Conyers, Lawrence B., Eileen G. Ernenwein, and Leigh-Ann Bedal
2002 Ground-penetrating radar (GPR) mapping as a method for planning excavation strategies, Petra, Jordan. E-tiquity Number 1 http://e-tiquity.saa.org/%7Eetiquity/title1.html

Cook, J. C.
1973 Radar exploration through rock in advance of mining. *Transactions of the Society of Mineral Engineering AIME* 254: 140–146.

1975 Radar transparencies of mines and tunnel rocks. *Geophysics* 40: 865–885.

Czarnowski, J., S. Geibler, and A. F. Kathage
  1996   Combined investigation of GPR and high precision real-time differential GPS. In *Proceedings of the Sixth International Conference on Ground Penetrating Radar*: 207–209. Department of Geoscience and Technology, Tohoku University, Sendai, Japan.

Daniels, David J.
  2004   GPR for landmine detection, an invited review paper. In *Proceedings of the Tenth International Conference on Ground Penetrating Radar: June 21–24, Delft, The Netherlands*, Evert Slob, Alex Yarovoy and Jan Rhebergen (editors). Delft University of Technology, The Netherlands and the Institute of Electrical and Electronics Engineers, Inc., Piscataway, New Jersey: 7–10.

Davenport, G. Clark
  2001a   *Where Is It? Searching for Buried Bodies and Hidden Evidence*. SportWork Press, Church Hill, Md.
  2001b   Remote sensing applications in forensic investigations. *Historical Archaeology* 35 (1): 87–100.

Davis, J. L., and A. P. Annan
  1989   Ground-penetrating radar for high-resolution mapping of soil and rock stratigraphy. *Geophysics* 37: 531–551.
  1992   Applications of ground penetrating radar to mining, groundwater, and geotechnical projects: selected case histories. In Pilon, J. S., Editor, *Ground Penetrating Radar*. Geological Survey of Canada, Paper 90-4: 49–56.

Davis, J. Les, J. Alan Heginbottom, A. Peter Annan, S. Rod Daniels, B. Peter Berdal, Tom Bergan, Kirsty E. Duncan, Peter K. Lewin, John S. Oxford, Noel Roberts, John J. Skehel, and Charles R. Smith
  2000   Ground-penetrating radar surveys to locate 1918 Spanish flu victims in permafrost. *Journal of Forensic Science* 45(1): 68–76.

Delaney Allan J., Steve A. Arcone, Allen O'Bannon, and John Wright
  2004   Crevasse detection with GPR across the Ross Ice Shelf, Antarctica. In *Proceedings of the Tenth International Conference on Ground Penetrating Radar: June 21–24, Delft, The Netherlands*, Evert Slob, Alex Yarovoy and Jan Rhebergen (editors). Delft University of Technology, The Netherlands and the Institute of Electrical and Electronics Engineers, Inc., Piscataway, New Jersey: 777–780.

Deng, Shikun, Zuo Zhengrong, and Huilian Wang
  1994   The application of ground penetrating radar to detection of shallow faults and caves. In *Proceedings of the Fifth International Conference on Ground Penetrating Radar*: 1115–1133. Waterloo Centre for Groundwater Research, Waterloo, Canada.

Dobrin, M. B.
1976   *Introduction to Geophysical Prospecting.* McGraw-Hill, New York.

Dolphin, L. T., R. L. Bollen, and G. N. Oetzel
1974   An underground electromagnetic sounder experiment. *Geophysics* 39: 49–55.

Doolittle, J. A.
1982   Characterizing soil map units with the ground-penetrating radar. *Soil Survey Horizons* 23: 3–10.

Doolittle, J. A., and M. E. Collins
1995   Use of soil information to determine application of ground penetrating radar. *Journal of Applied Geophysics* 33: 101–108.

Doolittle, James A., and Loris E. Asmussen
1992   Ten years of applications of ground penetrating radar by the United States Department of Agriculture. In *Fourth International Conference on Ground-Penetrating Radar: June 8–13, Rovaniemi, Finland*, Pauli Hanninen and Sini Autio (editors). Geological Survey of Finland Special Paper 16: 139–147.

Doolittle, James A., and W. Frank Miller
1991   Use of ground-penetrating radar techniques in archaeological investigations. In *Applications of Space-Age Technology in Anthropology. Conference Proceedings Nov. 28, 1990*, 2nd ed. NASA Science and Technology Laboratory, Stennis Space Center, Mississippi.

Engheta, N., C. H. Papas, and C. Elachi
1982   Radiation patterns of interfacial dipole antennas. *Radio Science* 17 (6): 1557–1566.

Fenner, Thomas J.
1992   Recent advances in subsurface interface radar technology. In *Fourth International Conference on Ground-Penetrating Radar: June 8–13, Rovaniemi, Finland*, Pauli Hanninen and Sini Autio (editors). Geological Survey of Finland Special Paper 16: 13–19.

Fischer, Peter M., S. G. W. Follin, P. Ulriksen
1980   Subsurface interface radar survey at Hala Sultan Tekke, Cyprus. In *Applications of Technical Devices in Archaeology*, Peter M. Fischer (editor). Studies in Mediterranean Archaeology 63: 48–51.

Fisher, E., G. A. McMechan, and A. P. Annan
1992   Acquisition and processing of wide-aperture ground-penetrating radar data. *Geophysics* 57: 495–504.

Fisher, S. C., R. R. Stewart, and H. M. Jol
  1994 Processing ground penetrating radar data. In *Proceedings of the Fifth International Conference on Ground Penetrating Radar*: 661–675. Waterloo Centre for Groundwater Research, Waterloo, Canada.

Freeland, Robert S., Ronald E. Yoder, and John T. Ammons
  1998 Mapping shallow underground features that influence site-specific agricultural production. *Journal of Applied Geophysics* 40: 19–27.

Fuchs, Michael, Milan Beres, and Flavio S. Anselmetti
  2004 Sedimentilogical studies of western Swiss lakes with high-resolution reflection seismic and amphibious GPR profiling. In *Proceedings of the Tenth International Conference on Ground Penetrating Radar: June 21–24, Delft, The Netherlands*, Evert Slob, Alex Yarovoy and Jan Rhebergen (editors). Delft University of Technology, The Netherlands and the Institute of Electrical and Electronics Engineers, Inc., Piscataway, New Jersey: 577–580.

Fullagar, P. K., and D. Livleybrooks
  1994 Trial of tunnel radar for cavity and ore detection in the Sudbury Mining Camp, Ontario. In *Proceedings of the Fifth International Conference on Ground Penetrating Radar*: 883–894. Waterloo Centre for Groundwater Research, Waterloo, Canada.

Gaffney, Chris, and John Gater
  2003 *Revealing the Buried Past: Geophysics for Archaeologists*. Tempus Publishing, Stroud, Gloucestershire, England.

Geophysical Survey Systems
  1987 *Operations Manual for Subsurface Interface Radar System-3*. Manual #MN83-728. Geophysical Survey Systems, North Salem, N.H.
  2000 *Radan for Windows: User's Manual MN43-162*. Geophysical Survey Systems, North Salem, N.H.

Gerber, Rolf, Peter Felix-Henningsen, Christina Salat, and Andreas Junge
  2004 Investigation of the GPR reflection pattern for shallow depths on a test site. In *Proceedings of the Tenth International Conference on Ground Penetrating Radar: June 21–24, Delft, The Netherlands*, Evert Slob, Alex Yarovoy and Jan Rhebergen (editors). Delft University of Technology, The Netherlands and the Institute of Electrical and Electronics Engineers, Inc., Piscataway, New Jersey: 275–278.

Goodman, D., and Y. Nishimura
  1993 A ground-radar view of Japanese burial mounds. *Antiquity* 67: 349–354.

Goodman, D., Y. Nishimura, and J. D. Rogers
　1995　GPR time-slices in archaeological prospection. *Archaeological Prospection* 2: 85–89.

Goodman, Dean
　1994　Ground-penetrating radar simulation in engineering and archaeology. *Geophysics* 59: 224–232.

　1996　Comparison of GPR time slices and archaeological excavations. In *Proceedings of the Sixth International Conference on Ground Penetrating Radar*: 77–82. Department of Geoscience and Technology, Tohoku University, Sendai, Japan.

Goodman, Dean, Yashushi Nishimura, Hiromichi Hongo, and Okita Maasaki
　1998　GPR Amplitude rendering in archaeology. In *Proceedings of the Seventh International Conference on Ground-penetrating Radar, May 27–30, 1998*. University of Kansas, Lawrence, Kansas, USA. Radar Systems and Remote Sensing Laboratory, University of Kansas: 91–92.

Goodman, Dean, Salvatore Piro, Yasushi Nishimura, Helen Patterson, and Vince Gaffney
　2004　Discovery of a 1st century A.D. Roman amphitheater and other structures at the Forum Novum by GPR. *Journal of Environmental and Engineering Geophysics* 9: 35–41.

Grasmueck, Mark
　1994　Application of seismic processing techniques to discontinuity mapping with ground-penetrating radar in crystalline rock of the Gotthard Massif, Switzerland. In *Proceedings of the Fifth International Conference on Ground Penetrating Radar*: 1135–1139. Waterloo Centre for Groundwater Research, Waterloo, Canada.

　1996　3D ground-penetrating radar applied to fracture imaging in gneiss. *Geophysics* 61: 1050–1064.

Grasmueck, Mark, Ralf Weger, and Heinrich Horstmeyer
　2004　Full resolution 3D GPR imaging for geoscience and archaeology. In *Proceedings of the Tenth International Conference on Ground Penetrating Radar: June 21–24, Delft, The Netherlands*, Evert Slob, Alex Yarovoy and Jan Rhebergen (editors). Delft University of Technology, The Netherlands and the Institute of Electrical and Electronics Engineers, Inc., Piscataway, New Jersey: 329–332.

Haeni, F. P., Marc L. Buursink, John E. Costa, Nick B. Melcher, Ralph T. Cheng, and William J. Plant
    2000   Ground-penetrating radar methods used in surface-water discharge measurements. In *Eighth International Conference on Ground-penetrating radar proceedings*. SPIE—The International Society for Optical Engineering, Bellingham, Wash.: 494–500.

Hatton, L., M. H. Worthington, and J. Makin
    1986   *Seismic Data Processing Theory and Practice*. Blackwell Scientific Publications, Boston, Massachusetts.

Heinz, J., and T. Aigner
    2003   Three-dimensional GPR analysis of various Quaternary gravel-bed braided river deposits (southwestern Germany). In *Ground Penetrating Radar in Sediments*, C. S. Bristow and H. M. Jol (editors). Geological Society Special Publication No. 211, The Geological Society, London 99–110.

Hildebrand, J. A., S. M. Wiggins, P. C. Heinkart, and L. B. Conyers
    2002   Comparison of seismic reflection and ground-penetrating radar imaging at the Controlled Archaeological Test Site, Champaign, Illinois. *Archaeological Prospection* 9: 9–21.

Hugenschmidt, J., M. N. Partl, and H. de Witte
    1998   GPR inspection of a mountain motorway in Switzerland. *Journal of Applied Geophysics* 40: 95–104.

Huggenberger, Peter, Edi Meier, and Milan Beres
    1994   Three-dimensional geometry of fluvial gravel deposits from GPR reflection patterns; a comparison of results of three different antenna frequencies. In *Proceedings of the Fifth International Conference on Ground Penetrating Radar*: 805–815. Waterloo Centre for Groundwater Research, Waterloo, Canada.

Imai, Tsuneo, Toshihiko Saskayama, and Takashi Kanemori
    1987   Use of ground-probing radar and resistivity surveys for archaeological investigations. *Geophysics* 52: 137–150.

Isaacson, John, R. Eric Hollinger, Darrell Gundrum, and Joyce Baird
    1999   A controlled archaeological test site facility in Illinois: Training and research in archaeogeophysics. *Journal of Field Archaeology* 26 (2): 227–236.

Ivashov, Sergey I., Sablin N. Vyacheslav, Anton P. Sheyko, and Igor A. Vasilev
    1998   GPR for detection and measurement of filled up excavations for forensic applications. In *Proceedings of the Seventh International Conference on Ground-penetrating Radar, May 27–30, 1998*. University of Kansas, Lawrence, Kansas,

USA. Radar Systems and Remote Sensing Laboratory, University of Kansas: 87–89.

Jackson, J. D.
1977 *Classical Electrodynamics*. Wiley, New York.

Johnson, Jay K.
2004 *Remote Sensing in Archaeology: A Cultural Resource Managers Guide*. University of Alabama Press, Tuscaloosa.

Johnson, R.W., R. Glaccum, and R. Wotasinski
1980 Application of ground penetrating radar to soil survey. *Soil Crop Science Society Proceedings* 39: 68–72.

Jol, Harry M., and Arlen Albrecht
2004 Searching for submerged lumber with ground penetrating radar: Rib Lake, Wisconsin, USA. In *Proceedings of the Tenth International Conference on Ground Penetrating Radar: June 21–24, Delft, The Netherlands*, Evert Slob, Alex Yarovoy and Jan Rhebergen (editors). Delft University of Technology, The Netherlands and the Institute of Electrical and Electronics Engineers, Inc., Piscataway, New Jersey: 601–604.

Jol, Harry M., and Charlie S. Bristow
2003 GPR in sediments: advice on data collection, basic processing and interpretation, a good practice guide. In *Ground Penetrating Radar in Sediments*, C. S. Bristow and H. M. Jol (editors). Geological Society Special Publication No. 211, The Geological Society, London 9–27.

Jol, H. M., and D. G. Smith
1992 Ground penetrating radar of northern lacustrine deltas. *Canadian Journal of Earth Science* 28: 1939–1947.

Jol, H. M., D. G. Smith, and R. A. Meyers
1996 Digital ground penetrating radar (GPR): A new geophysical tool for coastal barrier research (examples from the Atlantic, Gulf and Pacific Coasts U.S.A.). *Journal of Coastal Research* 12: 959–968.

Keller, George V.
1988 Rock and mineral properties. In *Applications in Electromagnetic Methods in Applied Geophysics*, edited by M. N. Nabighian: 13–24. Volume 1. Society of Exploration Geophysics. Tulsa, Oklahoma.

Kemerait, Robert C.
 1994 Ground penetrating radar considerations for optimizing the data collection scenario. In *Proceedings of the Fifth International Conference on Ground Penetrating Radar*: 761–775. Waterloo Centre for Groundwater Research, Waterloo, Canada.

Kenyon, Jeff L.
 1977 Ground-penetrating radar and its application to a historical archaeological site. *Historical Archaeology* 11: 48–55.

Kraus, J. D.
 1950 *Antennas*. McGraw-Hill, New York.

Kvamme, Kenneth L.
 2003 Geophysical surveys as landscape archaeology. *American Antiquity* 63 (3) 435–457.

LaFleche, P. T., J. P. Todoeschuck, O. G. Jensen, and A. S. Judge
 1991 Analysis of ground-penetrating radar data: Predictive deconvolution. *Canadian Geotechnical Journal* 28 (1): 134–139.

Lanz, Eva, Laura Jemi, Roger Muller, Alan Green, Andre Pugin, and Peter Huggenberger
 1994 Integrated studies of Swiss waste disposal sites: Results from georadar and other geophysical surveys. In *Proceedings of the Fifth International Conference on Ground Penetrating Radar*: 1261–1274. Waterloo Centre for Groundwater Research, Waterloo, Canada.

Leckebusch, J.
 2000 Two and three-dimensional ground-penetrating radar surveys across a medieval choir: A case study in archaeology. *Archaeological Prospection* 7: 189–200.
 2003 Ground-penetrating radar: A modern three-dimensional prospection method. *Archaeological Prospection* 10: 213–240.

Leckebusch, Jurg, and Ronald Peikert
 2001 Investigating the true resolution and three-dimensional capabilities of ground-penetrating radar data in archaeological surveys: Measurements in a sand box. *Archaeological Prospection* 8: 29–40.

Lehmann, Frank, and Alan G. Green
 1999 Semiautomated georadar data acquisition in three-dimensions. *Geophysics* 64 (3): 719–731.

2000 Topographic migration of georadar data: Implications for acquisition and processing. *Geophysics* 65 (3): 836–848.

Lehmann, Frank, Heinrich Hortsmeyer, Alan Green, and John Sexton
1996 Georadar data from the northern Sahara Desert: Problems and processing strategies. In *Proceedings of the Sixth International Conference on Ground Penetrating Radar*: 51–56. Department of Geoscience and Technology, Tohoku University, Sendai, Japan.

Lehmann, Frank, David E. Boerener, Klaus Holliger, and Alan G. Green
2000 Multicomponent georadar data: Some important implications for data acquisition and processing. *Geophysics* 65 (5): 1542–1552.

Leopold, Matthias, and Jorg Volkel
2003 GPR images of periglacial slope deposits beneath peatbogs in the Central European Highlands, Germany. In *Ground Penetrating Radar in Sediments*, C. S. Bristow and H. M. Jol (editors). Geological Society Special Publication No. 211, The Geological Society, London 181–189.

Loker, W. M.
1983 Recent geophysical explorations at Ceren. In *Archaeology and Volcanism in Central America*, Payson D. Sheets (editor): 254–274. University of Texas Press, Austin.

Lucius, Jeffrey H., and Michael H. Powers
2002 GPR data processing computer software for the PC. USGS Open File Report 02-166. Washington D.C.

Maijala, P.
1992 Application of some seismic data processing methods to ground penetrating radar data. In *Fourth International Conference on Ground-Penetrating Radar: June 8–13, Rovaniemi, Finland*. Pauli Hanninen and Sini Autio (editors). Geological Survey of Finland Special Paper 16: 103–110.

Malagodi, S., L. Orlando, and S. Piro
1994 Improvement of signal to noise ratio of ground penetrating radar using CMP acquisition and data processing. In *Proceedings of the Fifth International Conference on Ground Penetrating Radar*: 689–699. Waterloo Centre for Groundwater Research, Waterloo, Canada.

1996 Approaches to increase resolution of radar signal. In *Proceedings of the Sixth International Conference on Ground Penetrating Radar*: 283–288. Department of Geoscience and Technology, Tohoku University, Sendai, Japan.

Malagodi, S., L. Orlando, and F. Rosso
1996   Location of archaeological structures using GPR method: Three-dimensional data acquisition and radar signal processing. *Archaeological Prospection* 3: 13–23.

Martinaud, Michel, Michel Frappa, and Remy Chapoulie
2004   GPR signals for the understanding of the shape and filling of man-made underground masonry. In *Proceedings of the Tenth International Conference on Ground Penetrating Radar: June 21–24, Delft, The Netherlands*, Evert Slob, Alex Yarovoy and Jan Rhebergen (editors). Delft University of Technology, The Netherlands and the Institute of Electrical and Electronics Engineers, Inc., Piscataway, New Jersey: 439–442.

Marukawa, Yuzo, and Hiroyuki Kamei
1999   Estimation of the systematic error of three-component geomagnetic data using the ABIC method. *Archaeological Prospection* 6: 135–145.

McGeary, Susan, Julia F. Daly, and David E. Krantz
1998   High resolution imaging of Quaternary coastal stratigraphy using ground penetrating radar. In *Proceedings of the Seventh International Conference on Ground-penetrating Radar, May 27–30, 1989.* University of Kansas, Lawrence Kansas, USA. Radar Systems and Remote Sensing Laboratory, University of Kansas: 273–277.

Meats, C.
1996   An appraisal of the problems involved in three-dimensional ground-penetrating radar imaging of archaeological features. *Archaeometry* 38 (2): 359–379.

Milligan, Robert, and Malcolm Atkin
1993   The use of ground-probing radar within a digital environment on archaeological sites. In *Computing the Past: Computer Applications and Quantitative Methods in Archaeology*, Jens Andresen, Torsten Madsen, and Irwin Scollar (editors): 21–33. Aarhus University Press.

Moffat, D. L., and R. J. Puskar
1976   A subsurface electromagnetic pulse radar. *Geophysics* 41: 506–518.

Moran, Mark, Steve A. Arcone, Allen J. Delaney, and Roy Greenfield
1998   3-D migration/array processing using GPR data. In *Proceedings of the Seventh International Conference on Ground-penetrating Radar, May 27–30, 1998.* University of Kansas, Lawrence, Kansas, USA. Radar Systems and Remote Sensing Laboratory, University of Kansas: 225–228.

Neubauer, W., A. Seren Eder-Hinterleitner, S. and P. Melichar
   2002   Georadar in the Roman civil town Carnuntum, Austria: An approach for archaeological interpretation of GPR data. *Archaeological Prospection* 9: 135–156.

Neves, Fernando A., John A. Miller, and Mark S. Roulston
   1996   Source signature deconvolution of ground penetrating radar data. In *Proceedings of the Sixth International Conference on Ground Penetrating Radar:* 573–578. Department of Geoscience and Technology, Tohoku University, Sendai, Japan.

Nobes, David C.
   1999   Geophysical surveys of burial sites: A case study of the Oaro Urupa. *Geophysics* 64 (2): 357–367.

Noon, David A., Dennis Longstaff, and Richard J. Yelf
   1994   Advances in the development of step frequency ground penetrating radar. In *Proceedings of the Fifth International Conference on Ground Penetrating Radar:* 117–131. Waterloo Centre for Groundwater Research, Waterloo, Canada.

Olhoeft, G. R.
   1981   Electrical properties of rocks. In *Physical Properties of Rocks and Minerals,* Y. S. Touloukian, W. R. Judd, and R. F. Roy (editors): 257–330. McGraw-Hill, New York.

   1986   Electrical properties from 10-3 to 109 Hz-Physics and chemistry. In *Proceedings of the 2nd International Symposium of Physics and Chemistry of Porous Media,* J. R. Bananvar, J. Koplik, and K. W. Winkler (editors): 281–298. Schlumberger-Doll, Ridgefield, Conn.

   1994a Modeling out-of-plane scattering effects. In *Proceedings of the Fifth International Conference on Ground Penetrating Radar, June 12–16, Kitchener, Ontario, Canada*: 133–144. Waterloo Centre for Groundwater Research, Waterloo, Ontario, Canada.

   1994b Geophysical observations of geological, hydrological and geochemical heterogeneity. In *Proceedings of the Symposium on the Application of Geophysics to Engineering and Environmental Problems, Boston, MA, March 27–31, 1994*: 129–141.

   1998   Electrical, magnetic and geometric properties that determine ground penetrating radar performance. In *Proceedings of the 7th International Conference on Ground-Penetrating Radar:* 177–182. Waterloo Centre for Groundwater Research, Waterloo, Canada.

Olhoeft, G. R., and D. E. Capron
  1993 Laboratory measurements of the radio frequency electrical and magnetic properties of soils from near Yuma, Arizona. U.S. Geological Survey Open File Report 93-701. Washington, D.C.

Olson, C. G., and J. A. Doolittle
  1985 Geophysical techniques for reconnaissance investigation of soils and surficial deposits in mountainous terrain. *Soil Science Society of America Journal* 49: 1490–1498.

Ovenden, S. M.
  1994 Application of seismic refraction to archaeological prospecting. *Archaeological Prospection* 1 (1): 53–64.

Paniagua, Jesus, Mariano del Rio and Montana Rufo
  2004 Test site for the analysis of subsoil GPR signal propagation. In *Proceedings of the Tenth International Conference on Ground Penetrating Radar: June 21–24, Delft, The Netherlands,* Evert Slob, Alex Yarovoy and Jan Rhebergen (editors). Delft University of Technology, The Netherlands and the Institute of Electrical and Electronics Engineers, Inc., Piscataway, New Jersey: 751–754.

Pedley, H. M., and I. Hill.
  2003 The recognition of barrage and paludal tufa systems by GPR: case studies in the geometry and correlation of hidden Quaternary freshwater carbonate facies. In *Ground Penetrating Radar in Sediments,* C. S. Bristow and H. M. Jol (editors). Geological Society Special Publication No. 211, The Geological Society, London 207–223.

Pipan, M., I. Finetti, and L. Ferigo
  1996 Multi-fold GPR techniques with applications to high-resolution studies: Two case histories. *European Journal of Environmental and Engineering Geophysics* 1: 83–103.

Pipan, M., L. Baradello, E. Forte, A. Prizzon, and I. Finetti
  1999 2-D and 3-D processing and interpretation of multi-fold ground penetrating radar data: a case history from an archaeological site. *Journal of Applied Geophysics* 41: 271–292.

Piro, S., D. Goodman, and Y. Nishimura.
  2003 The study and characterization of Emperor Traiano's villa (Atopiani di Arcinazzo, Roma) using high-resolution integrated geophysical surveys. *Archaeological Prospection* 10: 1–25.

Piro, S., P. Maureillo, and F. Cammarano
  2000 Quantitative integration of geophysical methods for archaeological prospection. *Archaeological Prospection* 7: 203–213.

Powers, M. H., and G. R. Olhoeft
  1994 GPRMODV2: One dimensional full waveform forward modeling of dispersive ground penetrating radar data. U.S. Geological Survey Open File Report 95-58. Washington, D.C.
  1995 Waveform forward modeling of dispersive ground-penetrating radar. U.S. Geological Survey Open File Report 95-58, Washington, D.C.

Rees, Huw V., and Jonathan M. Glover
  1992 Digital enhancement of ground probing radar data. In Pilon, J., Editor. *Ground Penetrating Radar*. Geological Survey of Canada Paper 90-4: 187–192.

Reynolds, John M.
  1998 *An Introduction to Applied and Environmental Geophysics*. Wiley, New York.

Rhoades, J. D., P. A. C. Raats, and R. S. Prather
  1976 Effects of liquid-phase electrical conductivity: Water content and surface conductivity on bulk soil electrical conductivity. *Soil Science Society of America Journal* 40: 651–665.

Rojansky, Vladimir
  1979 *Electromagnetic Fields and Waves*. Dover, Mineola, N.Y.

Saaraenketo, Timo
  1998 Electrical properties of water in clay and silty soils. *Journal of Applied Geophysics* 40: 73–88.

Savvaidis, A., G. M. Tsokas, Y. Liritzis, and M. Apostolou
  1999 The location and mapping of ancient ruins on the Castle of Lefkas (Greece) by resistivity and GPR methods. *Archaeological Prospection* 6: 63–73.

Sellman, P. V., S. A. Arcone, and A. J. Delaney
  1983 Radar profiling of buried reflectors and the ground water table. *Cold Regions Research and Engineering Laboratory Report* 83-11: 1–10.

Sensors and Software
  1999 *Practical Processing of GPR Data*. Sensors and Software, Mississauga, Ontario.

Sheets, P. D.
  1992 *The Ceren Site: A Prehistoric Village Buried by Volcanic Ash in Central America*. Harcourt Brace Jovanovich, Fort Worth, Tex.

Sheets, P. D., W. M. Loker, H. A. W. Spetzler, and R. W. Ware
  1985 Geophysical exploration for ancient Maya housing at Ceren, El Salvador. *National Geographic Research Reports* 20: 645–656.

Sheriff, R. E.
  1984 *Encyclopedic Dictionary of Exploration Geophysics.* 2nd ed. Society of Exploration Geophysics, Tulsa, Okla.

Sheriff, R. E., and L. P. Geldart
  1985 *Exploration Seismology.* Cambridge University Press, New York.

Shih, S. F., and J. A. Doolittle
  1984 Using radar to investigate organic soil thickness in the Florida Everglades. *Soil Science Society of America Journal* 48: 651–656.

Shragge, Jeff, James Irving, and Brad Artman
  2004 Shot-profile migration of GPR data. In *Proceedings of the Tenth International Conference on Ground Penetrating Radar: June 21–24, Delft, The Netherlands,* Evert Slob, Alex Yarovoy and Jan Rhebergen (editors). Delft University of Technology, The Netherlands and the Institute of Electrical and Electronics Engineers, Inc., Piscataway, New Jersey: 337–339.

Simmons, G., D. W. Strangway, L. Bannister, R. Baker, D. Cubley, G. La Torraca, and R. Watts
  1972 The surface electrical properties experiment. In *Lunar Geophysics: Proceedings of a Conference at the Lunar Science Institute, Houston, Texas, 18–21 October 1971,* Z. Kopal and D. W. Strangway (editors). Reidel, Dordrecht: 258–271.

Smith, Derald G., and Harry M. Jol
  1995 Ground penetrating radar: Antenna frequencies and maximum probable depths of penetration in Quaternary sediments. *Journal of Applied Geophysics* 33: 93–100.

Stern, W.
  1929 Versuch einer elektrodynamischen Dickenmessung von Gletschereis. *Beitrage zur Geophysik* 23: 292–333.

Sternberg, Ben K., and James W. McGill
  1995 Archaeology studies in southern Arizona using ground penetrating radar. *Journal of Applied Geophysics* 33: 209–225.

Strongman, K. B.
  1992 Forensic applications of ground-penetrating radar. In *Ground-Penetrating Radar,* J. Pilon (editor), Geological Survey of Canada Paper 90-4: 203–211.

REFERENCES

Sun, J., and R. A. Young
    1995   Scattering in ground-penetrating radar data. *Geophysics* 6 (5): 1378–1385.

Tillard, Sylvie, and Jean-Claude Dubois
    1995   Analysis of GPR data: wave propagation velocity determination. *Journal of Applied Geophysics* 33: 77–91.

Todoeschuck, J. P., P. T. LaFleche, O. G. Jensen, A. S. Judge, and J. S. Pilon
    1992   Deconvolution of ground probing radar data. In *Ground-Penetrating Radar*, J. Pilon (editor). Geological Survey of Canada Paper 90-4: 227–230.

Tomizawa, Y., I. Arai, M. Hirose, T. Suzuki, and T. Ohhashi
    2000   Archaeological survey using pulse compression subsurface radar. *Archaeological Prospection* 7 (4): 241–247.

Turner, G.
    1992   Propagation deconvolution. In *Proceedings of the Fourth International Conference on Ground-Penetrating Radar June 8–13, Rovaniemi, Finland*. Geological Survey of Finland Special Paper 16.

Tyson, Peter
    1994   Noninvasive excavation. *Technology Review* February/March 1994: 20–21.

Urliksen, C. P. F.
    1992   Multistatic radar system-MRS. In *Proceedings of the Fourth International Conference on Ground-Penetrating Radar: June 8–13, Rovaniemi, Finland*, Pauli Hanninen and Sini Autio (editors). Geological Survey of Finland Special Paper 16: 57–63.

Valdes, Juan Antonio, and Jonathan Kaplan
    2000   Ground-penetrating radar at the Maya site of Kaminaljuyu, Guatemala. *Journal of Field Archaeology* 27 (3): 329–342.

Valle, S., L. Zanzi, H. Lentz, and H. M. Braun
    2000   Very high resolution radar imaging with a stepped frequency system. In *Proceedings of the Eighth International Conference on Ground Penetrating Radar*. Gold Coast Australia, May 23–26, 2000: 464–470.

Van Dam, R. I., and W. Schlager
    2000   Identifying causes of ground-penetrating radar reflections using time-domain reflectometry and sedimentological analysis. *Sedimentology* 47: 435–449.

Van Dam, Remke L., Elmer H. Van Den Berg, Sytze Van Heterren, C. Kasse, Jeroen A. M. Kenter, and Koos Groen
    2002   Influence of organic matter in soils on radar-wave reflection: sedimentological implications. *Journal of Sedimentary Research* 72: 341–352.

van Heteren, S., D. M. Fitzgerald, and P. S. McKinlay
    1994   Application of ground-penetrating radar in coastal stratigraphic studies. In *Proceedings of the Fifth International Conference on Ground Penetrating Radar*: 869–881. Waterloo Centre for Groundwater Research, Waterloo, Canada.

van Leusen, Martijn
    1998   Dowsing and archaeology. *Archaeological Prospection* 5: 123–138.

van Overmeeren, R. A.
    1994   High speed georadar data acquisition for groundwater exploration in the Netherlands. In *Proceedings of the Fifth International Conference on Ground Penetrating Radar*: 1057–1073. Waterloo Centre for Groundwater Research, Waterloo, Canada.
    1998   Radar facies of unconsolidated sediments in The Netherlands: A radar stratigraphy interpretation method for hydrogeology. *Journal of Applied Geophysics* 40: 1–18.

Vaughan, C. J.
    1986   Ground-penetrating radar surveys used in archaeological investigations. *Geophysics* 51 (3): 595–604.

Vickers, Roger S., and Lambert T. Dolphin
    1975   A communication on an archaeological radar experiment at Chaco Canyon, New Mexico. *MASCA Newsletter* 11 (1).

Vickers, Roger, Lambert Dolphin, and David Johnson
    1976   Archaeological investigations at Chaco Canyon using subsurface radar. In Lyons, Thomas R. (editor), *Remote Sensing Experiments in Cultural Resource Studies*. Chaco Center, USDI-NPS and the University of New Mexico: 81–101.

von Hippel, Arthur R.
    1954   *Dielectrics and Waves*. MIT Press, Cambridge, Mass.

Walden, A. T., and J. W. J. Hosken
    1985   An investigation of the spectral properties of primary reflection coefficients. *Geophysical Prospecting* 33: 400–435.

Walker, J. W., W. H. Hulse, and D. W. Eckart
    1973   Observations of the electrical conductivity of the tropical soils of western Puerto Rico. *Geological Society of America Bulletin* 84: 1743–1752.

Wensink, W. A.
  1993   Dielectric properties of wet soils in the frequency range 1–3000 MHz. *Geophysical Prospecting* 41: 671–696.

Woodward, John, Philip J. Ashworth, James L. Best, Gregory H. Sambrook Smith, and Christopher J. Simpson
  2003   The use and application of GPR in sandy fluvial environments: methodological considerations). In *Ground Penetrating Radar in Sediments*, C. S. Bristow and H. M. Jol (editors). Geological Society Special Publication No. 211, The Geological Society, London 127–142.

Worsfold, R. D., S. K. Parashar, and T. Perrott
  1986   Depth profiling of peat deposits with impulse radar. *Canadian Geotechnical Journal* 23: 142–154.

Wright, D. C., G. R. Olhoeft, and R. D. Watts
  1984   GPR studies on Cape Cod. In *Proceedings of the National Water Well Association Conference on Surface and Borehole Geophysical Methods*: 666–680. San Antonio, Tex.

Wright, D. L., and J. W. Lane Jr.
  1998   Mapping hydraulically permeable fractures using directional borehole radar and hole-to-hole tomography with a saline tracer. In *Symposium on the Application of Geophysics to Engineering and Environmental Problems, March 22–26, 1998, Chicago, Illinois, Proceedings*: 379–388. Wheat Ridge, Colo., Environmental and Engineering Geophysical Society.

Yelf, Richard
  2004   Where is true time zero? In *Proceedings of the Tenth International Conference on Ground Penetrating Radar: June 21–24, Delft, The Netherlands*, Evert Slob, Alex Yarovoy and Jan Rhebergen (editors). Delft University of Technology, The Netherlands and the Institute of Electrical and Electronics Engineers, Inc., Piscataway, New Jersey: 279–282.

Yilmaz, Oz
  2001   Seismic data analysis: Processing, inversion and interpretation of seismic data. Investigations in Geophysics Number 10, Society of Exploration Geophysicists, Tulsa, Okla.

Young, Roger A., and Sun Jingsheng
  1994   Recognition and removal of subsurface scattering in GPR data. In *Proceedings of the 5th International Conference on Ground Penetrating Radar*: 735–746. Waterloo Centre for Groundwater Research, Waterloo, Canada.

Yu, Haizhong, Xiaojian Ying, and Shi Yuansheng
 1996   The use of Fk techniques in GPR processing. In *Proceedings of the Sixth International Conference on Ground Penetrating Radar*: 595–600. Department of Geoscience and Technology, Tohoku University, Sendai, Japan.

Zeng, X., G. A. McMechan, J. Cai, and H. W. Chen
 1995   Comparison of ray and Fourier methods for modeling monostatic ground-penetrating radar profiles. *Geophysics* 60: 1727–1734.

Index

absorption, 23
acquisition: parameters, 9, 15; profiles, 13
adjustments: automatic, 84; manual, 84
air-waves: causes, 77; defined, 77; removal, 78
amplitude: analysis, 21; adjustments, 92; clipping, 93; enhancement, 50; rendering, 21
analog data, 31
antennas, 13, 23; bore-hole, 83; cables, 31, 83; coiled, 82; coupling, 28; frequencies, 66; housing, 27; low-frequency, 54; movement, 62; multiple, 21; ringing, 127; shielding, 77
attenuation, 23, 49, 50; adjustments with gains, 94; clay, 50; in soils, 50; range-gains applications, 92; with depth, 94

background noise: defined, 71; removal, 123, 124; sources, 71
background removal, pitfalls, 125
batteries, 82
bentonite clay, 52

bistatic mode, defined, 27
bow-tie effect in profiles, 139

cables, types, 83
caliche, 52
clay: bentonite, 52; conductivity, 52; mineralogy, 50; molecular structure, 50; structures, 169; types, 50; variables, 51
clipping, 93
clutter, 199; defined, 67; relationship to frequency, 67; relationship to object size, 67
CMP, 105
coefficient of reflectivity, 49
collection, continuous, 30; step, 30
common midpoint: calculations, 109; data collection, 107; depth of investigation, 109; displays, 108; three-dimensional analysis, 109
common-midpoint analysis, defined, 105
conduction, 23
conductivity, 50

197

conical propagation, 62–64
continuous collection, 30
control unit, 31; elements, 83
coupling, 28; changes, 70; anomalies, 71; related to ground, 70

data, analog, 31; downloading, 82; filtering, 50; interpretation, 133; raw, 13; storage, 82; transfer, 82
data collection, underground, 171; digitally, 5; manual, 5
data processing, 19, 15–16; grids, 11, 12; objectives, 120; pitfalls, 120, 131; post acquisition, 119; profiles, 11; programs, 119
deconvolution, predictive, 127; defined, 127
dielectric, defined, 50
dielectric constant, defined, 45; measured, 45
digital files, 19
direct-wave, defined, 90
dispersion, 49
display unit, 83
dog-leg paths, 136

electrical conductivity, 114
electromagnetic conductivity, 3
electromagnetic energy, propagation, 23–24
energy: depth of penetration, 86; penetration, 28, 54
excavation, planning, 5; projection, 13

features, elongated, 68
fiducial marks, 121; defined, 32; recording, 83
filter, vertical, 95
filtering, 50; examples, 97; F-k, 126
filters, 126; band-pass, 95; horizontal, 88; spatial, 88; vertical, 95

finite-difference, migration, 129
finite impulse response, defined, 126
fire pits, 169
F-k filtering, 126
focusing, 69, 73
footprint, 63
footprint-frequency relationship, 63
forensic applications, 17
forward modeling, 135
frequency, defined, 23; filtering, 96; footprint relationship, 63; and near-field effect, 76; relation to wavelength, 24; stepped, 82; units of, 24

gains, 91
geophysical surveys, successes, 8
global position systems (GPS), 28
grid, properties, 45; mapping, 147; transect orientation, 67
ground, acidity, 52; conductive, 52; coupling, 68; dry, 52; properties, 13; saturated, 52
groundwater table, 17

header information, 84
high-pass filters, defined, 95
Hilbert Transform, applications, 131; defined, 130
hip-wheel, 81
historical archaeology, 18
history of GPR, 16
horizontal banding, 123; removal, 130
horizontal smoothing, defined, 88
hydrology, 17
hyperbola analysis, 115
hyperbolas, 68; collapse, 129; computer fitting, 116–117; removal, 129

ice, 16
infinite impulse response, defined, 126

INDEX 199

interfaces, in grid, 54; sloping, 67
interference, 119
ionic movement, 52
iron-oxides, 53

kilns, 20, 169
Kirchoff Method, migration, 129

laboratory measurements, electrical properties, 113; for velocity analysis, 112; frequency dependence, 114; magnetic properties, 113; of RDP, 112
landscape, mapping, 145; ancient, 2, 20
low-pass, defined, 95

magnetic methods, 3, 6
magnetic permeability, 114; defined, 53
magnetic susceptibility, 3
manufacturers of geophysical devices, 6
manufacturers of systems, 81
memory, cards, 82
metal, 79
migration, applications, 128; finite-difference, 129; Kirchoff Method, 129; phase shift, 129; Stolt Method, 129
model, creation, 135
modeling, 19, 134
modeling, forward, 135
models, examples, 139–142; usage, 142–143
monostatic, defined, 27
mounds, burial, 18
multi-channel systems, 27
multiples, 119; production, 126; removal, 126

near-field effect, 75
near-field-frequency relationships, 76
near-field zone, defined, 76
network analyzer, 113

noise, 119; generation, 72; removal, 125; sources, 72

organic matter, 52

phase shift, migration, 129
pipes, 17
pit dwellings, 19, 169; model of, 144
point-sources, 68
power, usage, 82
predictive deconvolution, defined, 127
profiles, modeling, 134; axes, 32; examples, 120; rubber-sheeting, 121; scale correction, 120; topographic correction, 123
pulse, transmission rate, 89
pulses, 13; generation rate, 29; recording, 29; sampling, 29

radio spectrum, 24–25
range gains, applications, 92, 138; attenuation relationship, 91; defined, 91
ray tracing, defined, 136; examples, 139–142; theory, 136; variations, 137
RDP. *See* relative dielectric permittivity
reflected wave velocity methods, 100
reflection: causes, 25; coefficient, 49; from graves, 17; from walls, 17, 18; generation, 48–49; increase visibility, 129; nonhorizontal, 125; processed, 119; raw, 119
reflection profile, interpretation, 144; color, 121
reflection trace, defined, 26
refracted wave velocity methods, 100
relative dielectric permittivity (RDP): calculations of, 47–48; defined, 45; interfaces, 64; of materials, 45–47; measured, 45; relationship to focusing,

69; resistivity, 3; resolution, horizons, 64; spatial, 12
resolution, wavelength, 64
reverberations, 119
ring-down, 79
ringing, 127; filtering, 123
road pavements, 17
rock, properties, 25
rubber-sheet, defined, 121

salts, attenuation, 50; hydrous, 50
samples, collection, 87; definition, 87
scatter, defined, 73
sediment, properties, 25
seismic applications, modeling, 136
seismic methods, 4; processing, 126; applications to GPR, 126
set-up parameters, 84
set-ups, field, 9
shadow effect, 54
shadow zones, defined, 142
side scatter, defined, 67
Snell's Law, 68
snow, 119, 120; removal, 125
software programs, 6
soil: changes, 2; conditions, 172; attenuation in, 50; iron oxides in, 53; magnetic properties, 53; properties, 25; salts in, 50
stacking, defined, 88; reasons for, 89; variables, 89
static correction, defined, 122
step collection, 30
Stolt Method, migration, 129
stone tools, 169
stratigraphic analysis, velocity determination , 102
survey wheel, 28, 81; calibration, 81
synthetic model creation, 135–134; examples, 134

system, manufacturers, 81
system noise, removal, 123

three-dimensional reconstructions, 19
time zero, 90, 122
time-window, adjustments, 85; defined, 85; transmission relationship, 90; variables, 85–87
topographic correction, 123
trace, defined, 11; clipping 93; stacking, 88
trace spacing, 67
transect orientation, 67
transects, 2; collection, 28; density, 20
transillimination, data collection, 109, 111; defined, 105; examples, 110, 113; pitfalls, 111; set-up, 110
transmission pattern, 75
transmission rate, 89
trenches, 169
two-way travel time, defined, 11; plotted, 31

velocity, conversion for depth, 13
velocity, water saturation relationship, 101
velocity analysis, 20, 99; application to profiles, 122; direct waves, 105; general applications, 145; hyperbolas, 115; importance of, 116; laboratory tests, 112; pitfalls, 103–104; reflected waves, 100–103; refracted waves, 100; stratigraphy, 102; transillimination, 105
velocity pull-down, defined, 141
velocity pull-up, defined, 141
velocity variations, in modeling, 140
vertical filters, defined, 95
volcanic ash, 145

WARR, 105
water, relationship to velocity, 101

water saturation, 101; variables, 68
water table, velocity analysis, 104
wave, propagation, 23
waveform: clipping, 93; defined, 11; frequency variables, 88
wavelet, defined, 26
waves, EM, 50

wide-angle refraction and reflection velocity analysis: defined, 105; examples, 106
wiggle picking, 134
wiggle trace: defined, 120; examples, 121

zero time, 91, 122

About the Author

Lawrence B. Conyers is an associate professor of anthropology at the University of Denver, Colorado. He received a bachelor of science degree in geology from Oregon State University and a master of science degree from Arizona State University. He holds both M.A. and Ph.D. degrees in anthropology from the University of Colorado, Boulder. Before turning his attention to ground-penetrating radar and other near-surface geophysics for archaeological mapping, he spent seventeen years in petroleum exploration and development where he worked with seismic geophysical prospecting. His GPR research is conducted throughout the United States and at many sites throughout the world.